make your own
essential oils
& skincare products

second edition

Daniel Coaten

L I L I

First published in October 2006,
this second edition published in December 2013 by

Low-Impact Living Initiative
Redfield Community,
Winslow, Bucks, MK18 3LZ, UK
+44 (0)1296 714184

lili@lowimpact.org
www.lowimpact.org

Copyright © 2006, 2013 Daniel Coaten

ISBN 978-0-9566751-5-6

Design and Layout: Commercial Campaigns Ltd

Printed in Great Britain by:
Lightning Source UK Ltd, Milton Keynes

contents

illustrations

disclaimer

Anyone using the methods described in this book does so at his or her own risk. LILI, the author, and their associates assume no liability for any damage to persons or property caused by the use or misuse of the information contained herein.

The information and procedures detailed within this book are not for use by anyone unable to handle the potential risks associated with them.

about the author

Daniel Coaten is an experienced medical herbalist, auricular (ear) acupuncturist and aromatherapist. He studied western herbal medicine at Middlesex University (completing his degree in 2000), and went on to become a member of the National Institute of Medical Herbalists (NIMH). He has since been running his own complementary therapy clinic, and practices in Aylesbury, Buckinghamshire.

Whilst studying for his degree, he became interested in the practice of acupuncture. Subsequently, he completed a course in auricular acupuncture and became a member of the Substance Misuse Acupuncture Register (SMART). He continued his work within the substance misuse field, working with young people for a UK-based charity called Addaction, and is a member of the Federation of Drug and Alcohol Professionals (FDAP).

In 2005, Daniel completed his aromatherapy / holistic massage diploma through the Institute of Traditional Herbal Medicine and Aromatherapy (ITHMA) and is now a practicing member of the International Federation of Professional Aromatherapists (IFPA).

He now runs seminars on small-scale distillation of essential oils, herbal extraction techniques and making skin-care products, and is an international consultant to both the aromatherapy and herbal medicine manufacturing industries. He has written articles for *In Essence* (IFPA's journal), and the *International Journal of Clinical Aromatherapy*, and has been a guest speaker at NIMH and IFPA conferences.

Daniel's real passion is the manufacture of plant medicines. Over the years, he has gained the knowledge and skills necessary to produce many of the remedies required for his own clinic's dispensary. He has first-hand experience in growing medicinal herbs, as well as many of the techniques used in herbal and aromatic pharmacy. Daniel is also a member of the International Holistic Skin Care Producers Association (IHSA).

Daniel is director of 'Elixir Herbal Botanical Pharmacy' which produces herbal medicines, natural skin-care products and supplies equipment needed to extract, produce and research botanical-based remedies (www.elixirherbal.com).

His ethos is 'self sufficiency' and he enjoys nothing more than being able to connect with and follow the process of herbal pharmacy from seed to final product, witnessing the metamorphic magic that is plant medicine manufacture.

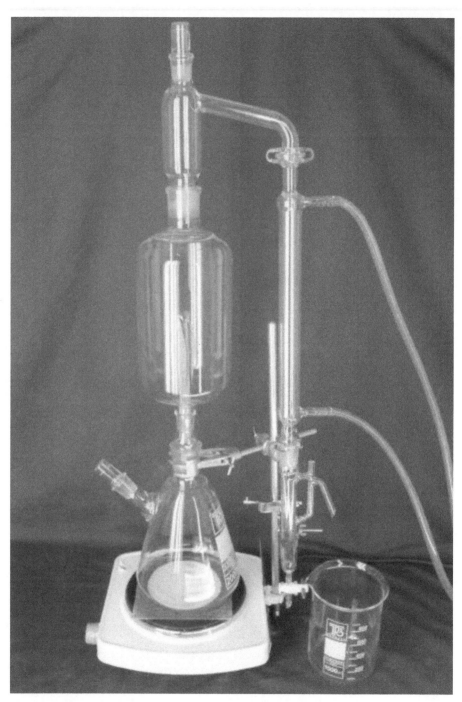

the standard distillation kit used for the procedures outlined in this book; for details of components see appendix 1

introduction

For many years now we have believed that the processes involved in the production of essential oils and skin-care products required expensive equipment, complicated procedures and extensive technical knowledge, and that it was not accessible for the average practitioner or layperson to perform at home.

Also, contrary to popular belief, the possession of small-scale distillation equipment, and the water distillation of essential oils / distillate waters is legal worldwide. This means that this type of distillation can be practiced by anyone, anywhere, without the fear of prosecution.

However, it is important to mention that if you were to use the equipment for the process of alcohol distillation without a government licence (in many countries including the UK), you would be breaking the law!

With so many companies now supplying a wide choice of essential oils, distillate waters and pre-made skin-care products, you may ask the question: 'why go to the bother of producing my own?'

Several good reasons come to mind. To start with there is the fact that there are still a large number of commercial producers of essential oils and skin-care products that adulterate their wares with synthetic 'aroma-chemicals' to meet the demanding standardisation needs of the flavour and fragrance industries. Many of these products may also contain harsh chemical preservatives (some of which may cause severe allergic reactions in some people!), and/or raw materials that are from ethically/ecologically unsound sources.

By producing your own supply of essential oils and skin-care products you can be sure of exactly how the products are produced, how fresh they are, their source, and that they have not been adulterated.

Secondly, for too long now aromatherapists have come to rely upon these commercial manufacturers to supply them with their 'tools of the trade', in the same way that doctors rely upon pharmaceutical companies.

By being able to produce your own supply of essential oils and skin-care products (or at least some of them) the average aromatherapy practitioner or lay person has the chance to effectively become self-sufficient if they so choose, in a similar way to medical herbalists growing their own herbs and producing their own medicines.

Lastly, the formulation and creation of your own supply of Aromatherapy and skin-care products is a fascinating and stimulating hobby in its own right, and one that can be thoroughly rewarding.

It is therefore the aim of the first part of this book to provide a simplified explanation to the processes involved in the distillation of essential oils / distillate waters and the extraction of aromatic compounds. This is then followed by an easy, fully illustrated, step-by-step guide to producing your own essential oils / distillate waters (using the example of 'the distillation kit') so that anyone can effectively and safely apply these techniques at home.

The second part of the book then gives the reader the opportunity to incorporate the products created from part one into further products by detailing all the equipment, raw materials, and procedures (again fully illustrated) needed to make your own macerated oils, herbal tinctures, gels, ointments, balms, creams, and lotions. Simplified recipes are given for each of these products as well as tips on customising your own formulas, in order that readers may tailor their end product to meet their own or their patients' needs.

Finally, further information about the raw materials used and products made from the above procedures can be found in the monographs, tables etc in the appendix section.

part 1:

aromatic extracts, essential oils & distillate waters

definitions

aromatic extracts

Aromatic products that are produced without the aid of the distillation process are not technically speaking EOs (although they may share many of the same properties as their respective EOs). These products are collectively known as aromatic extracts.

For more information on these products and how they are produced see 'methods used to produce aromatic extracts' on page 21.

essential oils.

An essential oil (EO) is a highly-concentrated aromatic substance, composed of volatile (i.e. evaporates easily) and relatively non-polar chemicals (i.e. not soluble in water) extracted from a plant source by means of distillation. Many are non - oily, though most have an 'oily feel' to them. They are soluble in fixed oils, alcohol, ether, and most organic (carbon containing) solvents and they are insoluble in water (being either lighter or denser than water - some sink, some float). However, a small volume of the non-polar compounds from the essential oil may disperse in water and impart its odour to the water (this is then known as EO water).

You should also be aware that the quality of many of the EOs sold around the world are not as high as it could be. Many of them are 'perfume grade' or are adulterated with other oils or synthetic chemicals. These second-rate oils do not have the therapeutic effect that 'clinical grade' oils do. Unfortunately, when low-quality oils are used, the therapeutic effects are also of a low quality, and low-quality results, in turn, undermine the reputation of aromatherapy as a whole.

It is therefore essential that practitioners be vigilant when purchasing commercially-produced EOs by asking for proof of quality control measures employed by their supplier, or alternatively, produce their own supply.

distillate waters

Distillate waters (DWs), also known as distillation or essential waters, are the waters that have been condensed, separated (from the EO) and collected from the process of distillation of herb material. DWs can be divided into the following three categories:

hydrosols / hydrolats

These waters are considered to be a by-product of steam distillation in the process of making EOs, usually on an industrial scale. They used to be discarded after the distillation process was complete, but more usually they are now sold on (or used in the process of cohobation - see page 78: 'a note on cohobation').

aromatic waters (AWs)

These are the primary product in the process of water (hydro or simple) distillation of an aromatic herb. In this case, it is the EOs that are considered to be the by-products. These waters have become enriched with both the EO, (more specifically EO water - EOw), and the water soluble volatiles (WSVs) that have been extracted from that herb material. The EOw component is made up of finely dispersed non- / semipolar molecules in a low concentration which contributes to the aromatic water's individual aroma. The WSVs in the solution help to give the aromatic water additional properties not possessed by the EO alone.

herbal waters (HWs)

Traditionally, it has been the aromatic herbs that have been chosen for distillation to produce EOs, AWs and hydrosols / hydrolats. Recently, however, some of the 'less aromatic' herbs have been distilled and researched for their medicinal properties. These are produced in exactly the same way as either hydrosols or AWs, but may contain less of the aromatic components (EOws) commonly associated with AWs and so are more correctly given the term 'herbal water'.

Note: there are products on the market produced by simply adding EOs to water. These are often referred to as perfumed waters, but can be sometimes found to be misleadingly labelled as aromatic or floral waters. These products are not true DWs, as they are not the products of distillation and do not contain the same compounds (such as the WSVs) or share the same properties as authentic DWs.

For more information see appendix 2: 'distillate water monographs'.

methods used to produce aromatic extracts

enfleurage

Some flowers, such as jasmine or rose, have such low content of aromatic compounds or are so delicate that heating them would destroy the blossoms before releasing their EOs. In such cases, an expensive and lengthy process called enfleurage is sometimes used to obtain their aromatic products.

In this process, the flower petals are placed on sheets of glass that are prepared by coating them with bland (having very little scent), warmed oils / fats (in particular solid fats such as lard / waxes), which absorb the flowers' aromatic essences. Once prepared, the sheets of glass are stacked one on top of another.

Every day (or sometimes every few hours), after the fat has absorbed as much of the essence as possible, the depleted petals are removed and replaced with fresh ones. This procedure continues until the wax / fat or oil becomes saturated with the pure aromatic essences (this is known as a 'pomade'). Alcohol is then added to this pomade and shaken over several hours to extract / separate the aromatic extract from the fatty base. The alcohol is then evaporated and only the aromatic extract remains (known as an 'absolute').

Note: once used to create treasured pomades and floral 'creams', it is by far the most laborious, expensive and time-consuming extraction method. Consequently, this method is rarely applied to the fragrant essence trade today (except in some unusual cases such as with tuberose), and so remains the oldest truly 'traditional' method of aromatic extraction.

infusion

This is a term that is commonly used to describe the process of making a 'tea' using hot water and herb material. In this case however, the term is used to describe the process where oil (such as animal fat or vegetable / nut oil) is used instead of water to perform the extraction.

Heating the oil helps to soften / break down the cellular structure of the plant material, allowing for improved extraction of the aromatic compounds into the fat / oil medium. The exhausted plant material is then removed from the hot fat by either centrifuging or straining, and replaced with new herb material as needed (again, like enfleurage, until the fat has been fully saturated with the aromatic essences).

If oils have been used in this process, the final product is called an 'infused oil' (IO), but if fats / waxes have been used, the final product is called a 'pomade' (as in enfleurage), and the aromatic extracts can be processed and collected in the same way as for that technique (see page 21).

expression

This is the method of extracting aromatic essences by cold-pressed expression (once commonly referred to as 'scarification'). It is used to obtain citrus fruit oils such as bergamot, grapefruit, lemon, mandarin, orange etc.

In this process, pressure is applied to the fruit which is rolled over a trough with sharp projections that penetrate the peel, piercing the tiny pouches and releasing the 'zest'. Then, either the peels are misted with a fine spray of water and the aromatic extract separated / collected, or the whole fruit is pressed to squeeze the juice from the pulp, and the aromatic extract is separated from the juice by centrifuging.

At one time the oil-rich rind of these fruits were squeezed by hand and collected into sponges. Some of the Mediterranean producers of these extracts still use this age-old method rather that relying on machines and the technology of centrifugal force to do the work.

Note: due to the enormous demand for citrus extracts, the quality available to the public today can vary greatly. Many are by-products of major juice manufacturers, who offer citrus extracts as a side commodity to their main business, after they have separated the fruit, pith and juice.

solvent extraction

Another, more modern method of extraction used on delicate plant material, is solvent extraction. This process is similar to infusion (mentioned above). However, in this case, a chemical solvent (such as hexane, ether, benzene or other petroleum-based compounds) is used to saturate the plant material and extract the aromatic compounds rather than fat / oil. The resultant semi-solid mixture of solvent and aromatic extract is known as a 'concrete', and will contain waxes and residues which must be filtered away and purified with the help of alcohol washing, freezing, and gentle vacuuming. The alcohol is then evaporated to obtain the pure aromatic extract (also known as an 'absolute').

The primary advantage of this method is that a low, uniform temperature range and low pressure is employed and maintained, allowing the flower's signature fragrance to be captured, which may otherwise be damaged / altered by heating. The resultant colours of the absolutes are very rich, and their consistencies range (depending on the filtering) from a thickened cream to resinous, and they may even crystallize or become semi-solid at lower temperatures.

Note: although more cost-effective than enfleurage or maceration, solvent extraction has its disadvantages. Residues of the solvent may remain in the absolute (especially that of ethyl alcohol, which can be present in the range of 1-2%), and some can cause unwanted side-effects. While absolutes or concretes may be fine for fragrances or perfumes, they are not especially desirable for natural skin care applications.

Some trees (such as benzoin, frankincense, and myrrh), exude aromatic 'tears' or sap that is too thick to use easily in aromatherapy. In these cases, a resin or absolute (which can be easier to use) can be obtained from the tears with the use of alcohol. You don't really want to be using hexane, benzene or chloroform as they are potentially dangerous (although scarily, still used in some food fragrances and flavourings, as they are cheaper). Alcohol is better, but some aromatherapists won't use absolutes at all.

carbon dioxide extraction

A more recent method that has been developed is that of the 'supercritical' carbon dioxide extraction. This process uses carbon dioxide (CO_2) under extremely high pressure to extract the aromatic essences. Plants are placed in a stainless steel tank, and as carbon dioxide is injected into the tank, pressure inside the tank builds. Under high pressure the carbon dioxide turns into a liquid and acts as a solvent to extract the aromatic compounds from the plants. When the pressure is decreased, the carbon dioxide returns to a gaseous state, leaving no residues behind.

Many CO_2 extractions have fresher, cleaner, and crisper aromas than the steam-distilled EOs, often smelling more similar to living plants (this is primarily due to the inert nature of the solvent), and containing some of the heavier / 'larger' components that would otherwise be absent in the corresponding EOs.

The CO_2 extracts available can be divided into two types:

total CO_2 extracts

These are normally thicker in consistency than the 'select' type of extract (often needing to be held under warm water to achieve a pourable consistency), and contain the extracted plant's oils, waxes, fats, resins, and colour.

select CO_2 extracts

These contain less solid plant material, and as a result tend to be easier to pour and use. They also tend to be slightly more expensive due to the extra processing needed to achieve this consistency.

- This extraction method uses lower temperatures than steam-distillation, making it gentler on the plants and their corresponding aromatic compounds (although there is some disagreement, as due to the acidic nature of CO_2, one could argue that it is disruptive to the chemistry of the resultant extracts).

It produces higher yields and makes some materials (especially gums and resins), easier to handle. Many aromatic extracts that cannot be produced by steam distillation could be obtainable with CO_2 extraction.

Scientific studies are now showing that CO_2 extraction produces aromatic extracts that are very potent and may have great therapeutic benefits, although there still needs to be a lot of research done on the safety of these extracts for aromatherapy and skin-care use.

phytonic extraction

This extraction employs a family of benign gaseous solvents known as non-chlorinated fluorocarbons (non-CFCs); namely 'R134a' (or refrigerant hydrofluorocarbon 134a - now called 'florasol'). Their development in the early 1980s to replace CFCs attracted the curiosity of the British microbiologist, Dr. Peter Wilde. Dr. Wilde brought this process into the spotlight of the aromatic extract manufacturing community, inventing the innovative extraction method of phytonic extraction in England.

The process is similar to that of CO_2 extraction, where the hydrofluorocarbon solvent is injected into a high-pressure sealed unit containing the aromatic plant material. The liquid gas that is formed under this high pressure is then released into another sealed chamber where it evaporates. The end result is a pure aromatic extract containing the most delicate aromatic components and a mere 2% of moisture. This technique has now been in development over the past 10 years; the resulting extracts were once called 'phytols', then 'phytosols' and are now referred to as 'florasols'.

butane extraction

This is a rather new method developed and used by French companies, producing a 'butaflor'. Again, this process is similar to that of CO_2 / phytonic extraction, except that this time butane is used as the solvent to extract the aromatic products. One of the advantages in the employment of butane is that the boiling point is low and therefore residual traces of it can be removed easily without using much heat or vacuuming. Unfortunately, the downside is that butane is caustic and highly flammable, making it difficult to work with.

distillation

history of distillation

The word distillation refers to 'dropping' or 'trickling' (according to its Latin roots), and is best described as the process of separating out a miscible combination of liquids into its individual components via their differing boiling points. It is an ancient technique which is commonly employed today for refining chemicals, alcohol, and fossil fuels, and for producing EOs / DWs.

In 1975 a perfectly preserved terracotta distillation apparatus from about 3000 BC was found in Pakistan.

In about 1810 B.C. in Mari, Mesopotamia, (the site of the present city of Tall al-Hariri, Syria), the perfumery of King Zimrilim employed this method to make hundreds of litres of balms, essences and incense from cedar, cypress, ginger and myrrh every month, using a huge vessel with a lid and a sluice.

Queen Cleopatra is also thought to have known about distillation, and is said to have given an account of the process in a text which is now lost.

In the first century it was mentioned by the Greek physician Pedanius Dioscurides (who travelled with the armies of the Roman emperors Nero and Vespasian) and at the turn of the second and third centuries by the Egyptian alchemist Zosimus of Panopolis near Thebes.

Distillation techniques were also developed in China between 300 and 600 AD.

However, it is thought that the first real advances in the distillation process were achieved by ancient Arabian alchemists, the most famous of whom was known in the West as Avicenna. In 900 AD he managed to extract the EO of roses by the then little-understood method of distillation. These alchemists produced most of the early texts on distillation, and Avicenna himself wrote many of them. It is Avicenna who is credited with inventing the cooling condenser, which he used to produce his DWs and EOs.

In 1000 AD, members of the Benedictine monastery of Salerno were credited with developing the modern still in this region.

Recently, detailed descriptions of distilling EOs have been found, written by the Spanish physician Anald de Villanova in the thirteenth century.

In medieval times, distillation was the alchemists' secret, and the equipment could be seen in every alchemical laboratory. Not until the early fifteenth century was the secret revealed, when Michele Savanrola of Italy described the method of extracting 'spiritus vini' from wine. In 1500, Hieronimus Brunschwygk's work 'Liber de Arte Destillandi' (Book of the Distilling Art) appeared in Strasbourg. During the rest of the century it was reprinted five times in Latin, and translated into Flemish (1517) and English (1527).

The simplest still (sometimes referred to as a 'retort') had a glass, copper, tin or ceramic flask, often mounted on a special lamp or burner, with a glass, metal or ceramic cap where drops of condensed vapour collected. The cap was beaked, that is, it funnelled to a tube leading down to a receptacle called the receiver. The flask, cap and beak together formed the 'alembic'. The parts were connected and sealed with 'lutum sapientiae' (solder of wisdom), a mixture of clay, powdered brick, egg white and horse manure.

The device was gradually improved and in 1526, Paracelsus used a water bath (called 'balneum mariae' by the alchemists) for the first time. This improvement prevented the flask from cracking whilst it was heating up, and also helped to stabilise the temperature of the liquid being distilled. The vapour cooling system was also improved, by running the tube through vessels of cold water, and in 1771, the German chemist Christian Ehrenfried Weigel invented an apparatus later wrongly named as the 'Liebig condenser', the forerunner of the condensing equipment of today. In it, the tube leading the distillate out of the still was 'jacketed' by another one flowing with water. Thus the alembic, a creature of the gloomy laboratories of the alchemists, gave birth to the complicated rectification devices of later times. During the same period, the alembic survived virtually unchanged in noblemen's households, right up to the mid-nineteenth century, where it was used to distil spirits and prepare aperitifs.

basic principles of distillation

There are four basic principles / phases involved in the process of EO / DW distillation: vaporisation; transportation; condensation; and separation. Each of these principles will be looked at in context using the following simplified description of the distillation process.

By way of example I will explain the process of water (hydro) distillation using a simple still known as a 'retort'. This is basically an enclosed container that has one small 'stoppered' hole at the top to allow the filling of the still and a small opening which extends into a tapered side 'arm'. This arm leads from the opening down to a smaller collecting container situated to the side of the retort (see fig 1: the retort, on page 29).

The first step is to soak the aromatic plant material with an adequate amount of water, then pour this mixture into the retort and apply the stopper. Heat is then applied to the bottom of the retort, so that the water / herb mix boils gently. The process of boiling produces steam, which helps to release the volatile components present within the herb material (the vaporisation phase). The steam rises towards the small opening in the retort and into the arm, carrying with it these volatile components (the transportation phase). As the steam / volatile compound mix enters the arm it starts to cool down, and turns from a gas back to a liquid (the condensation phase). At this point the room in which the distillation is taking place is usually filled with the aroma of the aromatic plant material!

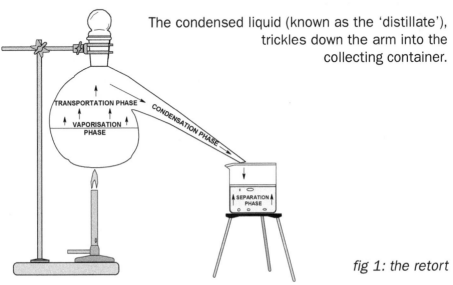

The condensed liquid (known as the 'distillate'), trickles down the arm into the collecting container.

fig 1: the retort

As the process continues, more and more liquid is collected and the distillate starts to separate out (the separation phase) into 2 distinct layers (in this example forming a thinner top layer). This top layer is the freshly made EO and can be skimmed off and stored, whilst the larger bottom layer is the DW (or more specifically AW in this case) which can also be stored for later use.

Of course, the example given above is distillation in a very simplified form, and may not necessarily be the most practical or efficient way of producing EOs etc. Even so, there is much evidence to suggest that this type of still was used extensively by our ancestors to produce aromatic products.

small-scale distillation in the 21st century

For years the equipment available for people who wanted to distil their own EOs etc on a small scale was bulky yet produced very small yields.

However, through trial and error and better understanding of the distillation process as a whole, there have been many modifications made to the still design to improve its efficiency.

Together with the availability of a greater variety of materials, this has led to major advances in the production of equipment for distillation on a small scale, therefore making this a more viable choice for those who want to produce their own personal stocks of EOs and DWs at home.

development of materials

Traditionally, many stills were made of copper. This material is readily available, easy to work with and is a good conductor of heat. The reactivity of this metal is thought to improve the quality of some distillation products (such as from rose, and bitter orange petals) although this also means that it can tarnish easily, making it harder to keep clean and use when a variety of EOs are to be produced. Therefore, a copper still is usually reserved for the distillation of one product only.

Another traditional material used is brass. It is a little more difficult to work with, but has the advantage of being more robust and resistant to knocks and dents, and doesn't tarnish so readily.

Stainless steel is hard wearing and very resistant to corrosion, so can be cleaned easily when used for multiple distillations. This is what a majority of the large-scale / commercial stills are made from.

However, over time, the rubber seals used in these types of stills can be eroded away by the steam and EOs, and may need to be replaced.

Personally I prefer the use of an all glass still. The development of Borosilicate (Pyrex®) glass has meant that a high-quality, corrosive-resistant still can be produced and sold relatively cheaply. The use of ground glass joints allows for easy setup and dismantling, for storage and cleaning purposes.

These joints are also interchangeable, giving freedom to the user to practice a variety of distillation and extraction techniques with ease, and overcome the problems associated with rubber seals as used with metal stills.

The equipment also looks attractive, and (more importantly) because the material is transparent, allows the user to observe the wonders of distillation before their very eyes (a big advantage over the other materials mentioned).

evolutions in design

Advances in still design have resulted in a more efficient process, leading to higher yields with a minimum of effort. Although there are a variety of designs employed, the basic parts to a still remain the same.

Below is a diagram of a 'Lab style' Pyrex® still setup (fig 2, below) along with a brief description of each of these parts, and how / why these parts have been modified.

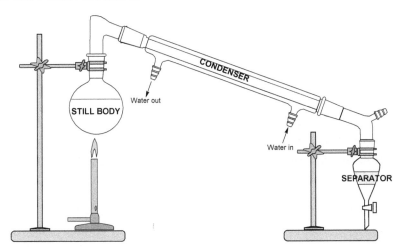

fig 2: 'lab style' Pyrex® still setup.

the still body (boiler)

This is the area of the still in which the plant material / water mix is placed. Heat is then applied to facilitate the vaporisation and transportation phases mentioned previously in 'the basic principles of distillation', page 29.

However, it was observed that in some cases this plant material was getting 'scorched' when it came in contact with the hot bottom / sides of the still, which resulted in an inferior product.

To remedy this it was suggested that the plant material be elevated off the bottom of the still and above the water via the use of a perforated grill, but which still allowed steam to access the plant material.

This modification in design started the era of 'steam distillation', and subsequent variations to this technique have now become the most common way to produce EOs commercially today.

the condenser

This is the area of the still that cools the vapour produced in the boiler, converting it back to a liquid form (distillate) and thus enabling the condensation phase.

A problem found when using the retort style still was that the yields of EOs produced were very low. It was noted that not all of the vapour produced was being converted back to its liquid state and so a lot of it was being lost to the atmosphere.

Placing the 'arm' part of the retort in or under running cold water increased the amount of vapour being converted to distillate, and yields were increased. Later, a 'jacketed' condenser (also known as a 'Liebig' condenser) was utilised, and subsequently led the way to many other modern designs based on this model (such as the coiled 'Graham' condenser). It was also found that by allowing the flow of water through the condenser jacket to be against that of the direction of the vapour increased the efficiency of the condenser still further (see figs 2-4).

Another important discovery involved the orientation of the condenser. Many traditional stills employ a condenser that is positioned at a 105º angle down from the boiler to the collection vessel (see fig 1-2); consequently there was still some vapour escaping. By simply repositioning the condenser vertically all of the vapour produced had nowhere else to go but down, and as a result even more of the vapour was cooled by the time it reached the separator, again increasing yields (see figs 3-4).

the separator

This is the area of the still where the distillate is collected and separates into its oil and water layers (the separation phase), before the two products are then siphoned off for storage.

The development of the 'Florentine flask' allowed for the products collected to be separated automatically for storage, eliminating the siphoning stage and enabling the user to simply tap off the chosen product as needed or at the end of the distillation process. Modern-day modifications to this design employ the use of a specialised separator known as a 'receiver / sep funnel' (see figs 3-4).

variations in technique

Along with the changes that were made in the design of the still there were also a number of variations made to the distillation technique. Again, many of the aforementioned principles remained the same, but it was found that by slightly adjusting the way the plant material was distilled you could accommodate for the different nature of plant material used (e.g. fresh or dried material, flower petals, powders etc).

In doing so, the producers of aromatic products could optimise their extraction yield and quality by meeting the individual needs of the plant materials used.

The main type of distillation obtainable with the distillation kit is (wet) steam distillation (see page 42). However, with the addition of a hydro-tube (sold separately), you can also do a hydro-distillation (see page 56). With this equipment you can achieve most types of small-scale EO and DW production. The employment of interchangeable ground glass joints allows the user to switch between (wet) steam and hydro - distillation with ease. However, with a little extra equipment (also sold separately) there are many other variations in extraction techniques obtainable with this kit. These will be discussed in more detail in the 'advanced techniques' section (page 79).

(wet) steam distillation

This process involves the elevation of plant material above the water level in the boiler, which allows the steam produced to permeate through this material when the still is heated. In this example a separate chamber altogether is used to house the herb material, whilst the lower flask acts as a boiler to produce the 'wet' steam (see fig 3: (wet) steam distillation setup, below). Note the extra sep funnel (see page 65).

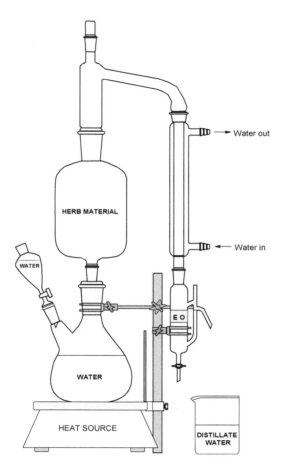

fig 3:(wet) steam distillation setup

The types of plant material best suited to this process are usually dried, hairy, absorbent herbs (e.g. dried lavender tops).

hydro-distillation

This process has already been described in 'the basic principles of distillation' section, page 29 (also see fig 4: hydro-distillation setup, below).

The types of plant material best suited to this process are flower petals (e.g. rose and bitter orange, which would otherwise disintegrate or form clumps), seeds, or ground / powdered herbs.

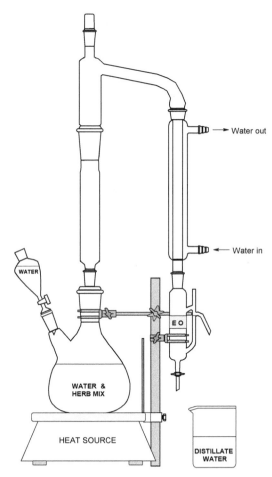

fig 4: hydro-distillation setup

basic laboratory health & safety

All workers using the distillation kit must assume a responsible attitude to their work. They must also try to avoid any careless, ignorant or rushed behaviour which may lead to an accident and possible injury to either themselves or to others. They should always pay due consideration to what is going on around them and be aware of any possible dangers arising from their procedures.

A majority of accidents are caused by attempts to obtain results in too great a hurry. Workers must therefore try to have an attentive, cautious and methodical approach to what they are doing.

personal protection

No worker should conduct procedures without a full-length protective lab coat (preferably white, so as to easily detect any spillages). Furthermore, all people present (i.e. including spectators) must wear safety spectacles / goggles at all times. In the event of an accident, these conventional safety spectacles / goggles may provide varying degrees of protection against flying fragments (depending upon type / make). However, they may provide very little protection against the splashing / spraying of hot liquids and gases. In the event of this happening, first aid should be confined to thorough irrigation of the eyes with clean water.

Note: if contact lenses are present they may restrict effective irrigation of the eye. However, their removal should only be performed by qualified medical staff.

personal conduct

Except in an emergency, shouting and running (or any other over-hurried activity), should be forbidden in and around the work area. Practical jokes and irresponsible behaviour are not acceptable. Smoking, eating and drinking in the work area should also be prohibited, and the worker should not conduct any procedures whilst tired or intoxicated with alcohol or drugs.

tidiness & cleanliness

The chosen work area should be both clean and tidy before any work is attempted, and floors must be kept well maintained to prevent slipping / tripping (i.e. they must be kept free from oil / water, and from any possible trip hazards). Passages around the area and near exits must not be blocked with equipment / furniture, and any spillages on the floor / work surface must be cleaned up immediately (this can be achieved by accessing wet and dry rags which can be kept close to hand).

Apparatus not being used immediately should be stored neatly and safely out of the way of the procedure, and dirty apparatus can be placed in a plastic bowl away from the working area until it can be cleaned and stored away.

All glassware should be meticulously clean and dry (both inside and out) before being used. It is advisable to get into the routine of cleaning all glass apparatus immediately after use, as the nature of any contaminants will then be known. Furthermore, the cleaning process may become increasingly difficult if dirty apparatus is allowed to stand for a considerable period (see 'care and maintenance of the distillation kit' (page 67) for more information on cleaning the glassware).

accident procedure

Every person working in the experimental area should ensure that he/she knows what to do in the event of an emergency / fire, where the exits and fire escapes are situated, and that there is free access to them. There should be first aid equipment (and people trained how to use it), fire extinguishers / blankets (and training), and a telephone for use in emergencies, all in (or close to) the work area. The checking of such equipment should be carried out by the proper authorities at regular intervals to conform to Health and Safety standards.

electrical safety

Many accidents are caused by electrical appliances malfunctioning and / or by their careless handling.

All new equipment should be suitably inspected for safety. Whether a plug needs to be fitted or the apparatus is already fitted with a plug, the appliance should always be inspected to ensure that:

- it is in good condition with no loose wires or connections
- it is properly earthed
- it is connected to the right type of plug, by a good quality cable that has the right insulation
- it is protected by a working fuse of the correct rating

Any loose / trailing electric cables should be avoided, and any items of equipment which have had any liquids spilled on them should not be used until they have been thoroughly cleaned and dried.

In the handling / setting up of electrical equipment, the worker must ensure that the apparatus is set up on a dry, heatproof work surface. It is essential that the apparatus is assembled first and only then plugged into the mains and switched on (also remember to switch off / allow to cool down before any attempts are made to move / adjust it).

glassware

Glass apparatus should always be carefully examined before use and any that is cracked, chipped or dirty should not be used. All apparatus and clean glassware not in use should be stored away and not be allowed to clutter up the work area.

planning experiments & recording results

It is always good practice to write up experiments done and results obtained in order to avoid repeating any mistakes, and to further your understanding of the processes involved in the experiment. Details such as the date and time of the experiment, the aim of the experiment, the apparatus used and how it was set up (simple diagrams can be useful here), the procedure (including details of amounts of substances used and when / how applied), the results (including how long the experiment took), and any thoughts on changes that could have been made to improve the procedure. This can be done easily if a 'lab journal' is kept and its contents habitually updated.

setting up the distillation kit

The first thing to do is to carefully unpack all the equipment from its box and lay it out onto a soft surface such as a bed. Then check that all the parts of the kit are present (see appendix 1: 'the distillation kit' (page 145), and that none of the equipment is damaged.

Next, carefully rinse all glassware parts with clean cold water and gently dry the outside with a clean tea towel or dishcloth.

The next thing to do is to attach the plastic hoses to the water inlet and outlet nipples of the 'West' condenser (see fig 5: connecting the hoses to the condenser, below). This is done by first lightly greasing the inlet / outlet nipples of the condenser with a small amount of the silicone grease supplied with the kit (a little goes a long way!). One end of one of the five-foot hoses is then lightly heated to soften the plastic (the easiest way to do this is to dip it into a beaker of freshly boiled water for approximately 10 seconds).

Note: it is advisable to wear eye protection and heat resistant / oven gloves when handling the beaker of boiling water.

When the plastic is soft enough it is then quickly, but steadily, pushed onto one of the nipples whilst gently turning until it is full on.

Note: it is advisable to wear hand and eye protection during this procedure, in case the condenser slips and / or breaks whilst fitting the hose.

After one hose is done then carry out exactly the same procedure for the other hose and nipple.

Once secured, there is really no need to remove the hoses again from the condenser (even when cleaning). If however, you do need to remove a hose then this can simply (and carefully) be cut free from the nipple. To reattach, use the same procedure as above again.

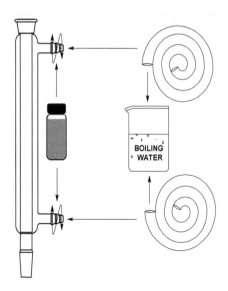

fig 5: connecting the hoses to the condenser

You are now ready to set up the distillation kit and can choose between doing a (wet) steam distillation or a hydro - distillation, depending on the type of herb material that you want to distil (see 'variations in technique', page 33).

The following two sections will give you a step-by-step guide on how to correctly set up the equipment and perform each of these types of distillation.

setting up the distillation kit for (wet) steam distillation

Note: extra equipment is needed to perform this distillation:

- a 'Phillips' screwdriver (appropriate size for the heat-shield screws and support rod secure screw).
- a small plastic funnel (to fit inlet neck of boiling flask).
- a long wooden chopstick.
- an electrical extension cable (as long as is needed for your particular setup).
- a large bowl (plastic) to hold the water / ice and water pump for the condenser.
- ice cubes / packs for the condenser water. These can be made beforehand and added as needed.

step 1

Find an appropriate place to set up the equipment (see 'basic lab health and safety', page 37). Position your hotplate on a heatproof work surface and have the rest of the equipment nearby. An appropriate sized 'Phillips' screwdriver is then used to secure the metal heat-shield to the hotplate with the screws provided (holes in the hotplate are already provided for this). See fig 6: step 1.

step 2

Insert the support rod (it can go either way up) into the support rod holder of the hotplate (see fig 7: step 2), and secure in place with the securing screw provided. The bottom of the rod should be touching the work surface.

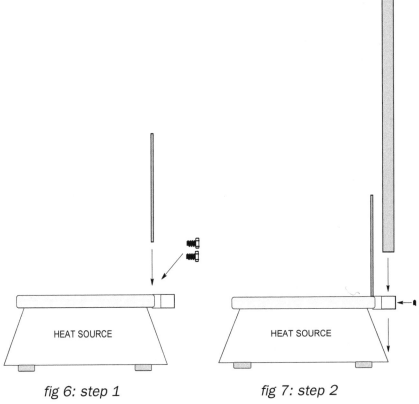

fig 6: step 1   fig 7: step 2

step 3

Position the wire-screen ceramic heat dissipater onto the hotplate ring (ceramic side up), and place the boiling flask on top (see fig 8: step 3).

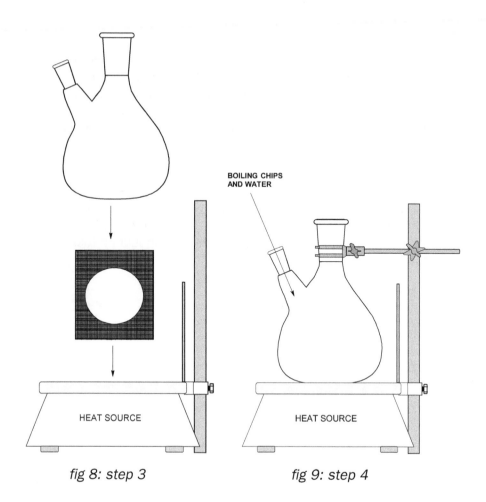

BOILING CHIPS
AND WATER

HEAT SOURCE

HEAT SOURCE

fig 8: step 3

fig 9: step 4

step 4

Secure the boiling flask in position using one of the laboratory clamps, after attaching it to the support rod (remember not to screw too tightly onto the glassware or it may break). See fig 9: step 4.

Now use a small plastic funnel to pour water into the boiling flask (filtered or distilled water is best). Do not fill the boiling flask more than ¾ full. Now add some boiling chips (3-4 pieces will do); these will help the water to boil evenly / steadily.

caution: if you forget to add these at the beginning, do not add them when the water is already hot - this is dangerous and results in a fountain of boiling water!

step 5

Place the stainless steel mesh screen filter (ready rolled into a tube) into the small (cone) end of the biomass flask (push in about half way). Apply a small amount of silicone grease to the socket adaptor for the boiling flask before inserting into the (centre) neck joint of the boiling flask. When in place rotate the biomass flask around several times clockwise and anti-clockwise to ensure an even coating of the grease to both joints (see fig 10: step 5).

Now apply silicone grease to the cone of the glass stopper that fits into the boiling flask's side arm. As above, rotate the stopper several times when in place to give an even coating of grease to the joint (see fig 10: step 5).

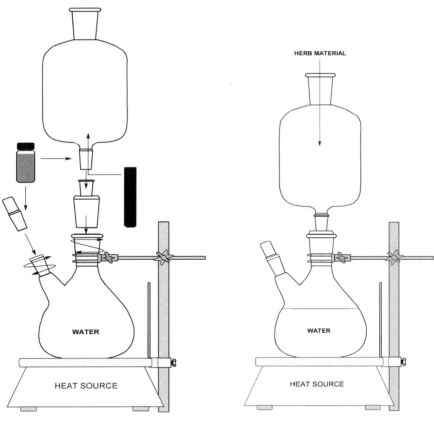

HERB MATERIAL

WATER

HEAT SOURCE

WATER

HEAT SOURCE

fig 10: step 5   *fig 11: step 6*

step 6

Pack the herb material into the biomass flask using a wooden chopstick (see fig 11: step 6). Remember that you will have to get the herb material out of the flask when cleaning it so cut up the herb material if it is too big or bulky, and try to pack the herb material evenly and not too tightly (remember that some dried herb material will swell during the distillation process).

Note: you can also use the hydro-distillation tube in place of the biomass flask here when distilling smaller amounts of herb material.

step 7

Make sure the joint of the biomass flask is free from herb material or dust (use a tissue to clean it). Then silicone grease the two cones of the still head and the cone of the stopper for the thermometer socket of the still head. Insert the larger cone into the joint of the biomass flask and put the stopper into the thermometer socket (again rotating the glassware to secure an even coat of grease to the joint). See fig 12: step 7.

step 8

Fill the receiver / separatory funnel with water (making sure that the stopcock is properly on and closed). This is done to make sure that none of the EO collected travels up the side arm of the receiver / separatory funnel and contaminates the DW collected.

Note: you will have to hold the glassware in place until the next step as the grease will not do this. It is a good idea to have the second laboratory clamp already lined up and in place for step 9 (fig 14).

Silicone grease is applied to the cone of the 'West' condenser, and the receiver / separatory funnel joint is fitted onto this (again rotating).

Note: the rubber tubing on the condenser is not shown on the diagram for ease of viewing but is still present.

The condenser joint is then fitted onto the ready-greased (small) cone of the still head (again rotating and applying a slight pressure up onto the cone). See fig 13: step 8.

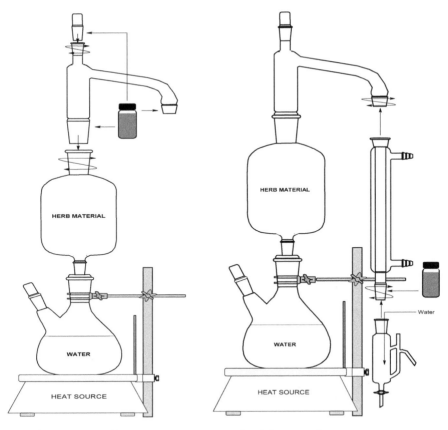

fig 12: step 7   fig 13: step 8

step 9

Fit the second laboratory clamp to the support rod and secure the receiver / separatory funnel and condenser in place (make sure there is a good seal at the still head / condenser joint and still a good seal at the still head / biomass flask joint).

Note: you may have to rotate the still head slightly to get the glassware in the right position to do this.

Use the 'Keck' clips provided to hold the joints together. The red Keck clip is for the spherical still head / condenser joint, and the green Keck clip is for the condenser / receiver / separatory funnel joint (see fig 14: step 9).

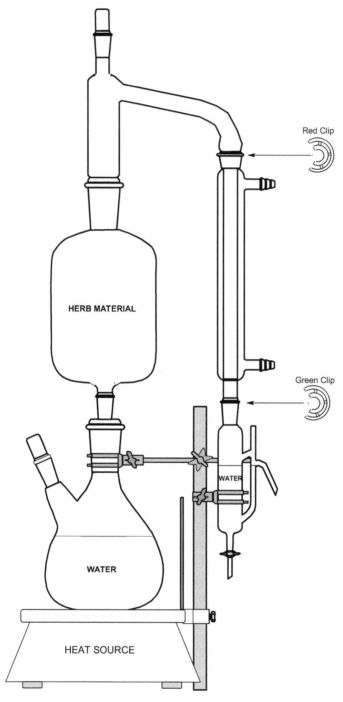

Red Clip

HERB MATERIAL

Green Clip

WATER

WATER

HEAT SOURCE

fig 14: step 9

step 10

Fill a bowl with water (not too full - remember you are going to add ice to it), and place near to where the apparatus has been set up.

Place the water pump in the water, and attach the rubber tubing leading to the water inlet of the condenser (i.e. the lower nipple of the condenser) to the outlet of the pump.

The rubber tubing leading from the water outlet of the condenser (i.e. the upper nipple of the condenser) should lead straight back into the bowl of water (see fig 15: step 10).
With dry hands plug the electrical extension cable into the nearest power point and turn on (away from where the water is!). Then plug the water pump into the extension socket.

You should slowly see the water from the bowl being pumped up the rubber tubing, into the jacket of the condenser, up and out of the top of the condenser, down the second length of rubber tubing, and back to the bowl again (see fig 15: step 10).

Note: the efficiency of the water pump may be impeded if any air bubbles are present. To remedy this, simply give the pump a little shake under the water to free the bubbles. The condenser's efficiency is also impeded by the presence of air bubbles. If this is the case, then the condenser may need to be detached from the apparatus and tilted so that the air bubbles are blown through, leaving just water present. The condenser is then re-attached in place with the water running through it.

Some ice is now added to the water in the bowl and topped up as needed.

step 11

Now orientate the collection flask so that it catches the overflow of DW from the sidearm of the receiver / separatory funnel (see fig 16: step 11).

step 12

Plug in and turn on the hotplate (by turning the heat dial on the front of the hotplate fully clockwise).

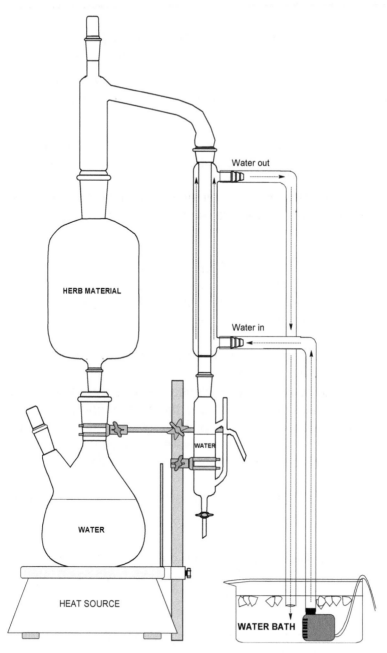

Labels within figure:
- Water out
- Water in
- HERB MATERIAL
- WATER
- WATER
- HEAT SOURCE
- WATER BATH

fig 15: step 10

caution: do not touch any part of the apparatus when it is operating as it will be extremely hot and could cause a burns injury.

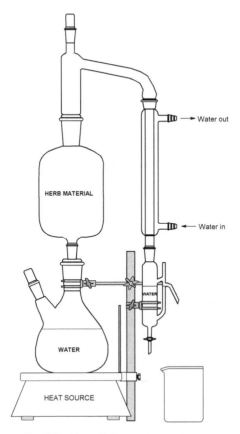

Water out

Water in

HERB MATERIAL

WATER

WATER

HEAT SOURCE

fig 16: step 11

Within 15-20 minutes you should start to see the water in the boiling flask begin to boil. A little while after this you will see the steam produced start to creep up into the biomass flask and saturate the herb material with steam. The steam will then move up into the still head and on into the condenser, where it will re-condense back into a liquid and drip into the receiver / separatory funnel. As more distillate is collected, the DW produced will overflow into the collection flask (via the receiver / separatory funnel sidearm) and the EO produced will remain in the receiver / separatory funnel for collection later.

Note: the temperature of the DW overflow should be no hotter than body temperature (as long as steam is NOT coming out of the overflow sidearm of the receiver / separatory funnel, you can check the temperature by allowing a few drops of DW to fall onto you finger to measure this). If the distillate is too hot or if steam is coming out of the overflow sidearm of the receiver / separatory funnel, then the hotplate needs to be turned down and more ice needs to be added to the condenser water bowl. A little fine tuning is needed with these until the right temperature of DW is obtained.

If the DW is too hot it is likely that the EO produced will be evaporating off, leading to decreased yields.

Note: remember to pour off the DW from the Collection flask into another glass container for storage before it overflows!

caution: do not allow the boiling flask to boil dry (i.e. all the water to boil off); at best it will be a nightmare to clean, and at worst it could be dangerous and shatter whilst hot! Always leave about 500ml in the boiling flask.

The whole process will take approximately 1½-2 hours to produce 1-1½ litres of DW and extract the EO from the herb material. The hotplate is then switched off and the apparatus is allowed to cool down completely before collection of the EO and cleaning of the glassware (you can leave the water running through the condenser until the water stops boiling in the boiling flask).

step 13

Remove the receiver / separatory funnel from the apparatus. If the EO produced is lighter than water (i.e. floats on the DW), then follow steps 13A-E below (figs 17-21).

If however, the EO produced is heavier than water (i.e. sinks below the DW), then follow steps 13F-I (figs 22 - 25).

procedure if EO is lighter than DW

step 13A

With the receiver / separatory funnel held over the empty collection flask, turn the stopcock anti-clockwise to open the valve (see fig 17: step 13A).

step 13B

Allow the DW to drain into the collection flask (see fig 18: step 13B). You can add this to the DW already collected.

step 13C

Turn the stopcock clockwise to close the valve when the EO reaches the bottom of the tube (you may need to slowly turn the stopcock before this time to control the flow of the EO). Try not to have any DW left in with the EO (see fig 19: step 13C).

If the EO seems cloudy, you can secure the receiver / separatory funnel upright using a laboratory clamp attached to the support rod, and leave for a couple of hours (or overnight) until it clears.

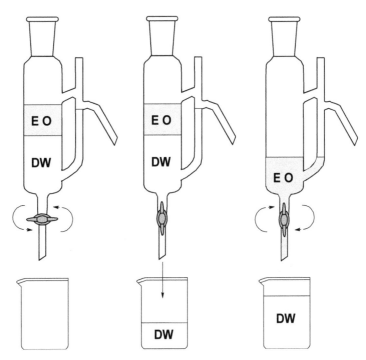

fig 17: step 13A   fig 18: step 13B   fig 19: step 13C

step 13D

Turn the stopcock anti-clockwise again to open the valve (see fig 20: step 13D).

step 13E

Allow the EO to drain into an appropriate glass container for storage (see fig 21: step 13E). You can now clean the receiver / separatory funnel.

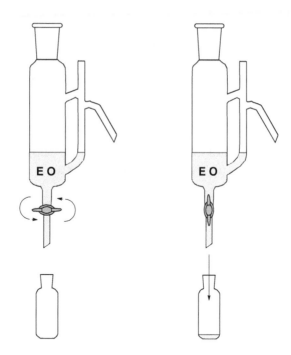

fig 20: step 13D   fig 21: step 13E

procedure if EO is heavier than DW

step 13F

With the receiver / separatory funnel held over an appropriate glass container turn the stopcock anti-clockwise to open the valve (see fig 22: step 13F).

step 13G

Allow the EO to drain into the appropriate glass container (see fig 23: step 13G).

step 13H

Turn the stopcock clockwise to close the valve just before the DW reaches the bottom of the tube (you may need to slowly turn the stopcock before this time to control the flow of the DW). It is OK to have a little bit of EO left in the receiver / separatory funnel to stop any DW getting into the EO storage container (see fig 24: step 13H).

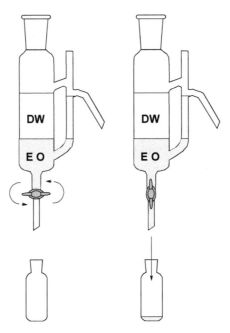

fig 22: step 13F *fig 23: step 13G*

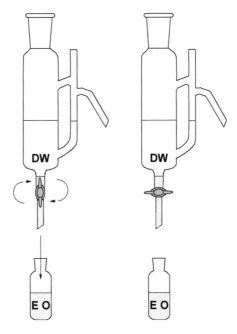

fig 24: step 13H *fig 25: step 13I*

step 13l

You can now discard the remaining DW and EO left in the receiver / separatory funnel (see fig 25: step 13l).

step 14

You can now clean out the glassware, rinse, dry, and store for use another day (see 'care and maintenance of the distillation kit', page 67, for more information on cleaning).

setting up the distillation kit for hydro-distillation

Note: extra equipment is needed to perform this distillation:

- hydro-tube (sold separately from the standard kit).
- a 'Phillips' screwdriver (appropriate size for the heat-shield screws and support rod secure screw).
- a small plastic funnel (to fit inlet neck of boiling flask).
- a long wooden chopstick.
- an electrical extension cable (as long as is needed for your particular setup).
- a large bowl (plastic) to hold the water / ice and water pump for the condenser.
- ice cubes / packs for the condenser water. These can be made beforehand and added as needed.

step 1

Find an appropriate place to set up the equipment (see 'basic lab health and safety', page 37). Position your hotplate on a heatproof work surface and have the rest of the equipment nearby. A 'Phillips' screwdriver is then used to secure the metal heat-shield to the hotplate with the screws provided (holes in the hotplate are already provided for this). See fig 26: step 1.

step 2

Insert the support rod (it can go either way up) into the support rod holder of the hotplate (see fig 27: step 2). Secure in place with the securing screw provided. The bottom of the rod should be touching the work surface.

step 3

Position the wire-screen ceramic heat dissipater onto the hotplate ring (ceramic side up), and place the boiling flask on top (see fig 28: step 3).

step 4

Secure the boiling flask in position using one of the laboratory clamps, after attaching it to the support rod (remember not to screw too tightly onto the glassware or it may break). See fig 29: step 4.

Now use a small plastic funnel to pour water and herb material into the boiling flask (filtered or distilled water is best). Do not fill the boiling flask more than ¾ full.

Boiling chips are not necessary in this type of distillation as the herb material present does the job of helping to initiate boiling.

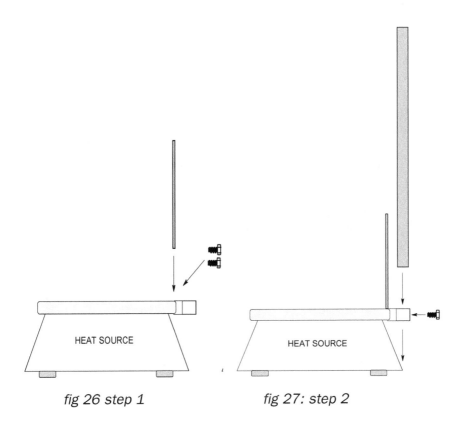

fig 26 step 1   fig 27: step 2

step 5

Apply a small amount of silicone grease to the cone of the hydro-distillation tube, and insert into the socket adaptor for the boiling flask; add silicone grease to the cone of the socket adaptor before inserting into the (centre) neck joint of the boiling flask. When in place, rotate the hydro-distillation tube around several times clockwise and anti-clockwise to ensure an even coating of the grease to the joint (see fig 30: step 5).

Now apply silicone grease to the cone of the glass stopper that fits into the boiling flask's side arm. As above, rotate the stopper several times when in place to give an even coating of grease to the joint (see fig 30: step 5).

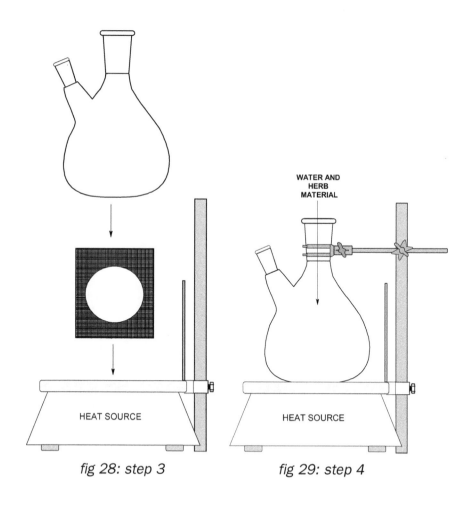

WATER AND
HERB
MATERIAL

HEAT SOURCE

HEAT SOURCE

fig 28: step 3   *fig 29: step 4*

step 6

Silicone grease the two cones of the still head, and insert the larger cone into the joint of the hydro-distillation tube (again rotating to secure an even coat of grease to the joint). See fig 31: step 6.

step 7

Fill the receiver / separatory funnel with water (making sure that the stopcock is properly on and closed). This is to make sure that none of the EO collected travels up the side arm of the receiver / separatory funnel and contaminates the DW collected.

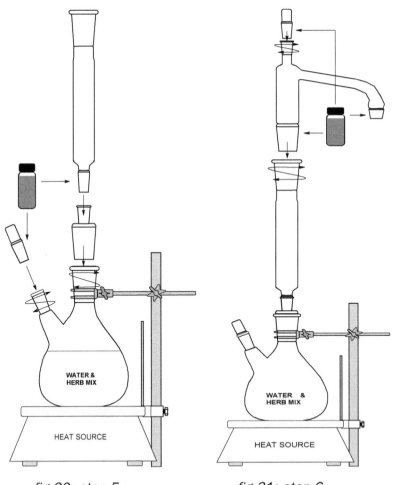

fig 30: step 5   fig 31: step 6

Note: you will have to hold the glassware in place until the next step as the grease will not do this. It is a good idea to have the second laboratory clamp already lined up and in place for step 8 (fig 33).

Silicone grease is applied to the cone of the 'West' condenser, and the receiver / separatory funnel joint is fitted onto this (again rotating).
Note: the rubber tubing on the condenser is not shown on the diagram for ease of viewing but is still present.

The condenser joint is then fitted onto the ready-greased (small) cone of the still head (again rotating and applying a slight pressure up onto the cone). See fig 32: step 7.

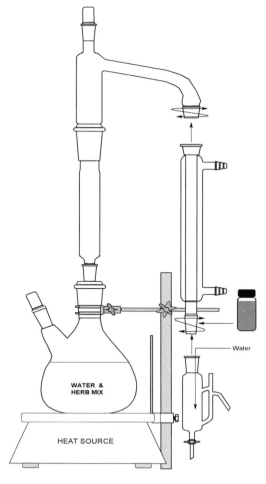

WATER & HERB MIX

Water

HEAT SOURCE

fig 32: step 7

step 8

Fit the second laboratory clamp to the support rod and secure the receiver / separatory funnel and condenser in place (make sure there is a good seal at the still head / condenser joint and still a good seal at the still head / hydro-distillation tube joint).

Note: you may have to rotate the still head slightly to get the glassware in the right position to do this.

Use the 'Keck' clips provided to hold the joints together. The red Keck clip is for the still head / condenser joint, and the green Keck clip is for the condenser / receiver / separatory funnel joint (see fig 33: step 8).

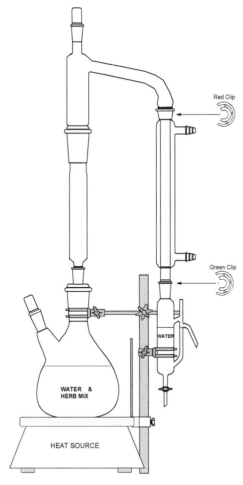

fig 33: step 8

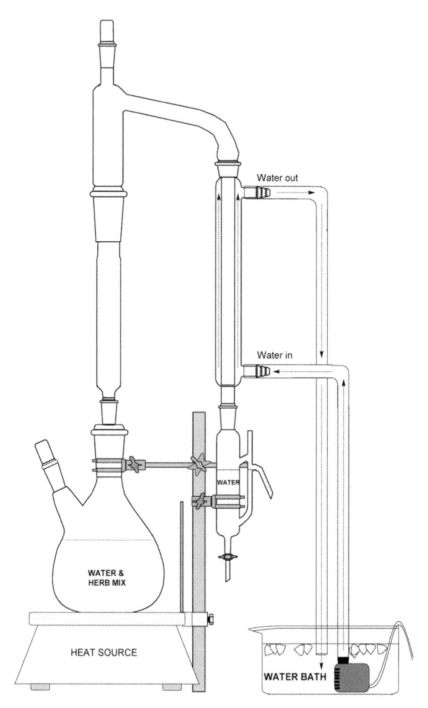

Water out

Water in

WATER

WATER &
HERB MIX

HEAT SOURCE

WATER BATH

fig 34: step 9

step 9

Fill a bowl with water (not too full - remember you are going to add ice to it), and place near to where the apparatus has been set up.

Place the water pump in the water, and attach the rubber tubing leading to the water inlet of the condenser (i.e. the lower nipple of the condenser) to the outlet of the pump.

The rubber tubing leading from the water outlet of the condenser (i.e. the upper nipple of the condenser) should lead straight back into the bowl of water (see fig 34: step 9).

With dry hands, plug the electrical extension cable into the nearest power point and turn on (away from where the water is!). Then plug the water pump into the extension socket.
You should slowly see the water from the bowl being pumped up the rubber tubing, into the jacket of the condenser, up and out of the top of the condenser, down the second length of rubber tubing, and back to the bowl again (see fig 34: step 9).

Note: the efficiency of the water pump may be impeded if any air bubbles are present. To remedy this, simply give the pump a little shake under the water to free the bubbles. The condenser's efficiency is also impeded by the presence of air bubbles. If this is the case, then the condenser may need to be detached from the apparatus and tilted so that the air bubbles are blown through, leaving just water present. The condenser is then re-attached in place with the water running through it.

Some ice is now added to the water in the bowl and topped up as needed.

step 10

Now orientate the collection flask so that it catches the overflow of DW from the sidearm of the receiver / separatory funnel (see fig 35: step 10).

step 11

Plug in and turn on the hotplate (by turning the heat dial on the front of the hotplate fully clockwise).

caution: do not touch any part of the apparatus when it is operating as it will be extremely hot and could cause a burns injury.

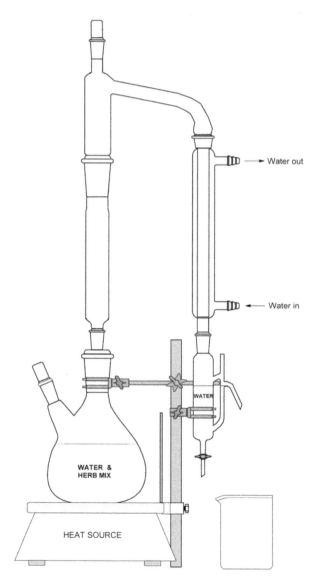

Water out

Water in

WATER

WATER &
HERB MIX

HEAT SOURCE

fig 35: step 10

Within 15-20 minutes you should start to see the water / herb mix in the boiling flask begin to boil (at this point turn the heat down to a 'gentle' boil). A little while after this you will see the steam produced creep up into the hydro-distillation tube, and then up into the still head and on into the condenser, where it will re-condense back into a liquid and drip into the receiver / separatory funnel. As more distillate is collected, the DW produced will overflow into the collection flask (via the receiver / separatory funnel sidearm) and the EO produced will remain in the receiver / separatory funnel for collection later.

Note: the temperature of the DW overflow should be no hotter than body temperature (as long as steam is NOT coming out of the overflow sidearm of the receiver / separatory funnel, you can check the temperature by allowing a few drops of DW to fall onto you finger to measure this). If the distillate is too hot or if steam is coming out of the overflow sidearm of the receiver / separatory funnel then the hotplate needs to be turned down and more ice needs to be added to the condenser water bowl. A little fine tuning is needed with these until the right temperature of DW is obtained.
If the DW is too hot it is likely that the EO produced will be evaporating off, leading to decreased yields.

Note: remember to pour off the DW from the collection flask into another glass container for storage before it overflows!

caution: do not allow the boiling flask to boil dry (i.e. all the water to boil off); at best it will be a nightmare to clean, and at worst it could be dangerous and shatter whilst hot! Always leave enough water for the herb mix to be still fluid to ensure ease of cleaning.

The whole process will take approximately 1-1½ hours to produce ½-1 litre of DW and extract the EO from the herb material. The hotplate is then switched off and the apparatus is allowed to cool down completely before collection of the EO and cleaning of the glassware (you can leave the water running through the condenser until the water stops boiling in the boiling flask).

steps 12–13

These steps are the same as for steps 13-14 (see pages 52-56 and figs 17-25) of the (wet) steam distillation section.

sep funnel reservoir

In order to extend the distillation time when distilling larger amounts of herb material (or if the material is especially rich in aromatic compounds) you can use a 500ml Pyrex® separatory funnel (regular type, not the one supplied with the distillation kit) as a water reservoir (see figs 3 & 4).

The equipment is set up as usual, except this time, at step 5, you fit the sep funnel to the side arm of the boiling flask in place of the stopper. You can then fill the sep funnel with water (with the tap closed!), so that you can add more water to the boiling flask without having to stop the distillation process. The sep funnel can then be refilled and used as needed to keep the boiling flask from boiling dry. This is especially useful when conducting hydro-distillations to keep the water level high in the boiling flask, therefore protecting the herb from scorching and possibly ruining your batch.

Note: try to keep an eye on the water level in the boiling flask, and add more water from the sep funnel before it gets too low. You should also try to add only a little at a time as a gentle trickle. In this way you can avoid cooling the boiling flask water too much, which can cause temperature fluctuations that can slow down or momentarily stop the distillation process.

care & maintenance of the distillation kit

cleaning

caution: be careful with the glassware - you do not want to find out how easily it can break!

You can use a chopstick or a pair of medical forceps to assist in getting the herb material out of the biomass flask. Shaking it in a plastic dustbin also works well to remove finer plant material (remember to hold on tight to the glassware!).

BE CAREFUL not to clunk the glass against the sink. It might be useful to line the sink with bubble wrap or something similar - most glassware breakages occur during cleaning! If the flasks bump into a hard object like a tile counter or porcelain sink, they may develop little tiny 'star' cracks. These are dangerous but they can be repaired. However, if unnoticed, they can easily cause a flask to crack or break apart at the most inconvenient time, like when it's full of boiling water!

When cleaning the glassware you can usually just swirl a lot of soapy water around in the parts (I like to use eco-friendly washing-up liquid), and use a bristle brush to clean everything (these can easily be bent into shape to facilitate this).

Mineral deposits (i.e. scale) that build up in the boiling flask can quickly be removed with a swirl of strong vinegar or dilute hydrochloric acid. A good soak with a limescale remover will probably work just as well. (BE CAREFUL with these chemicals - they are dangerous!) If the residue is particularly resinous, you could try using a little vodka in the rinse to try to dissolve this. An alternative and really efficient cleaning product is 'Decon 90'. This is a cleaning product specially designed for use with laboratory glassware, and is biodegradable and phosphate free as well. It can even remove charred remains from a flask!

The glassware is all Pyrex®, so it can also be put in a dishwasher (apart from the condenser, as it has the rubber hoses attached). It's a good idea to first wipe as much of the silicone grease out as you can with a paper towel. Otherwise just treat it as any other glass item and wash it as you will.

It is not necessary to remove the water hoses from the condenser (it's a lot easier to just leave these on for storage). To clean the condenser you can simply pour a little soapy water inside the centre tube, plug the ends with your fingers, and give it a good shake, rinsing with fresh cold water afterwards.

care of ground glass joints

Great care must be taken to ensure that all ground glass surfaces are free from grit and dust; during storage a strip of tissue paper can be inserted into joints and stopcocks to prevent jamming (also known as seizing, freezing, fusing or sticking). It is advisable, in order to reduce the danger of sticking, to apply a slight smear of silicone grease around the upper part of each ground joint (remember, a little goes a long way, so add a bit at a time with your finger). When greasing stopcocks, only the outer parts of the plug should be smeared with lubricant (care must be taken to avoid the lubricant getting into the bore of the plug and blocking it up).

Provided that you only use joints that fit well, and that the ground surfaces are suitably lubricated and parted after use (while still warm), sticking will rarely occur.

If, however, a ground glass joint should seize up the following suggestions may be found to be useful:

caution: care should be taken to avoid injury from breaking / shattering / broken glass - do not use excessive pressure on the glass, and wear suitable safety eyewear and gloves:

Sit the joint in a vertical position and apply a layer of glycerine or penetrating oil to the upper surface. The glycerine should slowly penetrate into the joint, thus permitting the separation of the ground surfaces.

If this procedure is unsuccessful, direct a stream of hot air from a blower (e.g. hairdryer) on to the outer surface of the joint for a few seconds and gently separate the cone from the socket with a slight twisting action.

Gentle tapping on the edge of a wooden bench can sometimes be helpful in aiding this procedure.

One last piece of advice: do not use ground glass joints that are damaged - this may result in other joints becoming damaged or even possible injury to yourself.

storage & shelf-life

Most of the EOs and DWs you will produce are probably not going to be used up straight after they are made. Therefore one of the most important things to decide is how you are going to store them for future use and yet make sure their quality remains high. If the products aren't stored properly they may not only lose their aesthetic appeal, but may also lose their therapeutic value or even become too hazardous to use.

To do this you need to consider the following factors which can affect your product's quality / shelf life:

degradation

It is a natural fact of life that all organic matter (both finished and raw products) degrades over time - some quicker than others. In the case of EOs and DWs this process is accelerated because of either prolonged storage or poor storage conditions.

storage conditions

The agents of degradation are oxygen, heat, light, and in some cases water.

The main means of degradation is via oxygen (which is called oxidation); this is made worse by the presence of both heat and light. Oxidation occurs in all EOs and DWs; however it seems to affect products that are rich in monoterpenes (such as pine and citrus products) more readily. Extra care is needed when storing these products, as they usually have the shortest shelf life.

Some EOs can also be affected by the presence of moisture. An example is *Lavandula angustifolia* (true lavender) EO, which contains linalyl acetate. If not properly dried when made, or if moisture is allowed to mix with it, the linalyl acetate is converted to linalool and plant acids resulting in the EO smelling off and becoming unusable.

To reduce the effect of these factors on your products and to limit their degradation, the following is advised:

- store the products in airtight containers that are either opaque or darkly coloured (I prefer using glass containers with corrosion-resistant lids).
- store the containers in a cool, dark, dry place.
- when opening the containers, do so for short periods of time only, and remember to replace the lids securely (some companies even flush out any oxygen in their containers with nitrogen gas, but this is unnecessary for the small-scale producer).
- use the smallest container possible for the amount of product present in order to reduce the amount of 'head space' (air space) above the product. As the product is used up, transfer to smaller containers as necessary.
- 'drying agents' (such as sodium or magnesium sulphate) are safe to use and can help make sure your EOs are dried properly. To do this, just add a small amount of the drying agent powder to the EO. The drying agent will soak up any water present without dissolving itself (it forms clumps). You then decant the EO off into its final storage container.

contamination

When producing your own EOs and DWs, contamination can fall into three main areas: contamination of starting material, contamination from organisms, and contamination from other products.

contamination of starting material

It is important to know where your starting materials (herb materials) for your products have come from and (if possible) how they have been grown. If you have grown the starting material yourself from seed then you should already know this; however, if you have wild crafted or bought your starting material from a supplier then do not be afraid to ask questions about it (i.e. 'how was it grown?', 'where was it grown?', 'how was it harvested?', 'when was it harvested?' etc). This is very important, because your product is only as good as the starting material it was exacted from.

contamination from organisms

This occurs more with DWs than EOs (mostly because EOs are in a too concentrated form for most organisms to survive).

The most common types of organisms that can be present are bacteria and fungi (especially yeasts and moulds), and usually occur if the utensils, equipment, and storage containers used are not properly cleaned /

sterilised before use. Some DWs however (especially HWs), are just prone to infection, and this is probably due to their having lower levels of EOws present that would otherwise help give protection against potential infections.

The following advice is therefore given to help prevent this type of contamination:

- bottle up your products as soon as possible after they have been made.
- make sure all equipment / glassware that comes in contact with the product is clean / sterilised (use of a pressure cooker is recommended).
- make up smaller batches of product and use them up as soon as possible.
- throw out any product that looks like it has been contaminated, and do not return any unused product back to its container.
- use the product for external application only. EOs should not be given to take internally unless you are specially trained in this field. Also, there is still very little research that has been done on DWs to warrant its safe use internally (see appendix 2: 'distillate water monographs' on page 147).
- some people use alcohol to try to preserve their DWs and extend their shelf life (the resulting mixture must be at least 25% alcohol to be effective as a preservative). However, before doing this, you need to decide how you are going to use your DW - as the addition of alcohol may affect this (e.g. if you are going to use it to make creams etc). I don't personally recommend this, as the alcohol can disrupt the delicate scent of the DW.
- in the case of DWs, only dispense small quantities from your storage container; use a spray / atomiser to apply the DW when used on its own (or in combination with other DWs); and if you are going to mix DWs together then only do so in small batches that will be used up quickly.
- if you have produced enough EO from a distillation (and it is lighter than the DW) you could add a little of this to the DW before it is stored. This would form a layer above the DW and help protect it from contamination. Remember that you will need to siphon off the DW without getting any EO with it (you can use a syringe to do this). This is not suitable for all DWs (see storage conditions, page 71)
- it is a good idea to store your DWs in a clean fridge to increase their shelf life and reduce the potential for contamination.

filtration and quality control of distillate waters

To be absolutely sure that your freshly-produced DWs are as free from organism contamination as possible it is recommended that they are filtered as soon as they are produced and before they are stored / aged. For small, personal batches of DWs you can use a simple water filter (e.g. a 'Brita'). In this instance you need to thoroughly clean and air dry the water filter and make sure that you use a new filter cartridge (remember to soak the filter in a quantity of the DW to be filtered before use) for each different DW you wish to filter.

However, if you are producing larger amounts then I recommend a 'Katadyn pocket microfilter'. This is more expensive than the filter mentioned above but is well worth the money if you plan to make a lot of DWs. This hand-powered pump can filter approximately half a litre a minute and has a ceramic design that can filter particles down to 0.2 microns; it is impregnated with colloidal silver (which can filter out or kill most common types of bacteria and viruses). Again it is recommended that the filter is dismantled, cleaned and air dried between filtering batches of DWs.

Note: of course all of this is a waste of time if the storage bottles used are not adequately cleaned and sterilised.

Another area of importance here is quality control. What is the shelf-life of the DW you have produced? And how can you tell when it is going off?

The first way of telling if a DW has passed its use-by date is by smelling it. If it starts to smell funny (possibly a little vinegary) then it is highly likely that it has started to go off due to a bacterial infection.

Another way of checking if the DW may have turned is by simply looking at it. It is common for many DWs to start to produce blooms (i.e. wispy, cloud-like particles that float in the DW) when they go off. This is usually due to a fungal infestation and simply filtering the DW through filter paper is usually not enough to make this problem go away.

The tests above can be useful but usually indicate late stages of infestation. What would be preferable is a way to closely monitor your DWs in order to detect an infestation before it ruins you batches of DWs. Fortunately, research in this area has shown that you can monitor the shelf-life of a DW by measuring its pH with universal indicator paper (or preferably an electronic pH meter). DWs produced from different herbs may display different starting pHs due to the differences in chemical constituents present. Even so it has been shown that during most common infections of DWs the pH usually starts to fall (i.e. become more acidic) due to the presence and activity of bacteria. You can test the pH of a sample of your DW as soon as it is produced (after filtering it), and with regular routine checks (every 2 months is recommended) the pH can be monitored, and deviations from the original pH (usually a change of more than pH 0.5) can be noted.

If a DW does start to show a lowered pH level then if it is caught in time it may be re-filtered using the Katadyn filter (the Brita filter will not work here). In most cases, when the filtered DW's pH is tested it will be back to its original value again. This technique can be successfully employed a few times to extend the shelf-life of a DW. However, these products do not last forever so at some point the DW will have to be discarded.

Note: for more information on filtration and monitoring of DW pH (including a table listing optimal pH ranges for commonly-produced DWs), please refer to *Hydrosols: the Next Aromatherapy*, by Suzanne Catty (see *resources* section).

contamination from other products

Again I must stress the importance of using properly cleaned / sterilized equipment / glassware. If you are doing a lot of distillations it is easy to contaminate one batch of product with the residues of another, so it is worth purchasing and using a pressure cooker or autoclave.

It is also important to think about the types of containers you use (either for transport or storage). You need to remember that both the EOs and the DWs contain substances that can be very corrosive to plastics (which may result in some of the plastic dissolving and contaminating your product) and that may even be reactive to some metals. It is therefore my preference to use all glass (amber coloured) storage containers with corrosion-resistant caps / dispensers.

breathing and ageing of products

After producing some DWs and EOs you may notice that they don't really smell as you would expect. A prime example of this is *Lavandula spp.* (lavender species). Often the DW and EO obtained directly after distillation smells a little 'off' (some people have said that it smells a bit like cabbage water). This is usually because of the presence of nitrogenous compounds that have been transferred in the process of distillation. Many commercial distillers solve this problem by simply allowing the EO and DW open to the atmosphere to 'breathe' after being separated. This process may last from several hours to overnight depending on the products, and allows most of the nitrogenous compounds to be released as gases, thereby improving the products' aroma.

In some cases (again with *Lavandula spp.*) the DW may need to be stored for several months or even a year for it to display its optimum aroma (in a process similar to that of ageing wines), and may indeed continue to improve with further aging if kept under favourable conditions.

labelling

One last thing to consider is labelling your product. This may seem obvious but it can easily be missed if you do a number of distillations in one go, resulting in having to guess which product is in which container. It is also important from a safety point of view (e.g. you should always write 'keep out of reach of children' and 'for external use only' on your product containers).

I find the following useful to include: the botanical and common name of herb material used, the amount of herb material used, the part of the plant that was used, whether the herb material was fresh or dried, the type of distillation used, the date the product was made (with possibly the date the product should be used by if applicable), a batch number, and who it was made by (if in a team).

herbs suitable for small-scale production

Below is a short list of some of the herbs I have distilled (including the type of distillation I used) to give you some idea of the sort of yields that can be expected from using the standard distillation kit (i.e. a 2-litre biomass flask), along with a hydro-tube and sep funnel reservoir.

herb material	distillation	EO yield	DW yield
Chamomilla recutita German chamomile (dried flowers)	(wet) steam	2ml	1.5 litres
Elettaria cardamomum cardamom (dried crushed seed pods)	hydro	15ml	1 litre
Eucalyptus globulus - blue gum eucalyptus (dried leaf)	(wet) steam	12ml	1.5 litres
Foeniculum vulgare - fennel (dried seed)	hydro	8ml	1 litre
Laurus nobilis - bay leaf (dried leaf)	(wet) steam	12ml	1.5 litres
Lavandula angustifolia - lavender (dried tops)	(wet) steam	8ml	1.5 litres
Melissa officinalis - lemon balm (dried herb)	(wet) steam	trace	1.5 litres
Mentha piperita - peppermint (dried herb)	(wet) steam	10ml	1.5 litres
Rosa damascena - rose (fresh petals)	hydro	Trace	1 litre
Rosmarinus officinalis - rosemary (dried tops)	(wet) steam	10ml	1.5 litres
Syzygium aromaticum - clove (dried buds)	hydro	18ml	1 litre

Note: obviously the EO content of a plant can change from sample to sample (depending on where and how it was grown / harvested etc). The samples presented here are a mixture of herbs I have grown myself and herbs I have bought from certified organic suppliers.

caution: by all means experiment with different herbs / quantities. However, be EXTREMELY CAREFUL if producing EOs / extracts of plants that are not already used therapeutically or have had little or no toxicological research done on them!

PLEASE DO NOT USE PLANTS THAT ARE ENDANGERED! This is not only illegal but also very selfish and irresponsible.

a note on cohobation

This technique is applied to the process of hydro-distillation and can be used to try to increase the EO yields of some types of herb material distilled (such as rose petals).

The process involves doing a hydro-distillation as normal. However, at the end of the distillation, when the apparatus is still set up and has been left to cool down (and the water in the boiling flask has stopped boiling), the stopper is carefully removed from the boiling flask's side neck (only when cool enough to do so without getting burnt!). Any DW produced is then poured into the boiling flask to join the already distilled herb material via a plastic funnel. The stopper is then replaced and the hotplate is turned on again to do another hydro-distillation.

This procedure can be repeated as many times as needed depending upon the herb material being distilled and its EO content. It should be noted here that the resulting DW from this process is usually of poor quality because it has been distilled many times, and is therefore not recommended for therapeutic use.

advanced techniques

reflux & the production of infused oils

This is similar to the infusion technique mentioned on page 21. In this case a condenser is used to ensure that the aromatic compounds extracted do not evaporate off, and refined vegetable oil is used so as not to impart its own odour onto the final product. The resulting product is known as an 'infused oil' (IO).

extra materials required

- a small plastic funnel (to fit the inlet neck of the boiling flask)
- a stirrer (preferably glass or stainless steel so it can be sterilised) - to stir the oil / herb mix
- an electrical extension cable (as long as is needed for your particular setup)
- a large bowl (plastic) to hold the water / ice and water pump for the condenser
- ice cubes / packs for the condenser water; these can be made beforehand and added as needed
- a saucepan of water (appropriate size to fit boiling flask / hotplate) - to use for the water bath
- a sheet of muslin cloth - to filter the oil / herb mix
- a glass container - clean / sterilised and big enough to house the filtered oil

procedure

A refined vegetable oil of your choice (the type of oil used depends upon what properties are required in the final product) is poured into the clean, dry boiling flask - no more than half full. Herb material (usually dried or powdered) is then added a little at a time to the oil (use a plastic funnel) whilst stirring to create a uniform mix (use a sterilised stirrer to do this). Do not try to add too much herb material to the oil at this stage as you will need the mix to remain fluid to be able to pour it out of the boiling flask when finished.

The apparatus is then set up as in fig 36: the reflux setup, below. Remember to grease the joints, set up the electrical extension cable (if needed) for the hotplate and water pump, and set up the bowl of ice / water for the condenser.

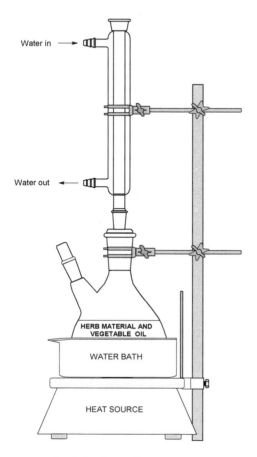

Water in →

Water out ←

HERB MATERIAL AND
VEGETABLE OIL

WATER BATH

HEAT SOURCE

fig 36: the reflux setup

Note: the flow of water through the condenser jacket is the reverse of that used in the previous distillations. However, it is still flowing against the direction of the vapour produced.

You will also see that a water bath is used in this setup (a simple cooking saucepan can be used); this is to provide an even heat to the oil / herb mix and ensures that the mix is not heated above 100°C, which may burn the oil and ruin the product.

caution: remember that the water in the water bath will boil and produce steam, so wear protective goggles and gloves, and be careful not to get burnt! Also, do not fill the water bath too full (as it may boil over), or allow it to boil dry!

The hotplate is then turned on full, and once the water in the water bath is boiling, the heat may be turned down to a simmer. The procedure continues for 2 hours and then the hotplate is turned off and the apparatus allowed to cool down completely (the condenser pump is left running until the apparatus has cooled - add more ice to the condenser water as needed).

When cool enough, the oil / herb mix is then poured out of the boiling flask and filtered through a muslin cloth into another glass container in order to remove as much of the herb material as possible (repeat this step if necessary). The resulting oil is then poured back into the boiling flask. More / new herb material is then added to the oil, and the above procedure is repeated.

The procedure is repeated as many times as is needed to produce the concentration of extract required (please note that most vegetable oils have a limit to the amount of times they can be reheated before they start to spoil - usually 4-5 times is maximum).

Note: if using fresh herb material you may find that a watery green liquid is present at the bottom of the oil at the end of the process. This liquid must be separated out and thrown away; otherwise it will go off and quickly spoil the IO (decant or use sep funnel).

The resulting IO can then be further filtered (see section on MO (page 84), then stored (preferably in a tinted glass bottle); it can be used as is, or be incorporated into ointments, creams etc. The advantage of employing this process rather than use the distillation technique is that more of the larger, heavier compounds will be present in the final extract that would otherwise be absent from an EO. These compounds often give the resultant extract an aroma that much more closely resembles that of the original herb material, and may also help to further the therapeutic actions of the extract. The shelf-life will be approx. 1 year, but 0.5-1% vitamin E oil can be added to extend the shelf-life a further year.

Note: for more information on IOs, please see appendix 3: 'macerated / infused oil monographs'.

(dry) steam distillation

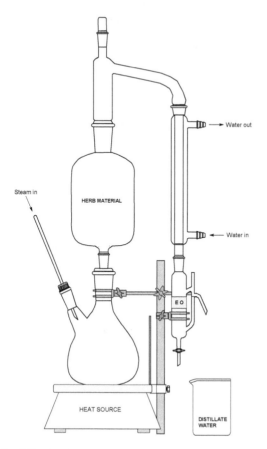

fig 37: the (dry) steam distillation setup

This process involves the production of steam in a separate (external) generator (such as a pressure cooker) under high pressure, producing a superheated 'dry' steam. This steam is injected into the still containing the plant material, and is the only source of heat used in this distillation (see fig 37: the (dry) steam distillation setup, above).

The types of plant material best suited to this process are fresh / wilted, non-absorbent, non-hairy herbs (e.g. wilted peppermint leaf), as the drying action of the steam helps to prevent reflux and 'flooding' of the still (which may reduce yields and quality of end product).

Note: this is an advanced distillation technique requiring extra equipment to perform.

caution: working with pressurised steam can be very dangerous. This technique is not recommended for the beginner and must only be performed by those who are skilled and confident in working with this medium (and wearing the appropriate safety gear).

steam-injected hydro-distillation

This process is similar to the last example, except this time water is added to the plant material in the boiling flask (as in the case of hydro - distillation) before steam is then injected into the water and herb mix (see fig 38: The steam injected Hydro - distillation setup, below). The still may also be heated directly (via the hotplate) as well as from the steam.

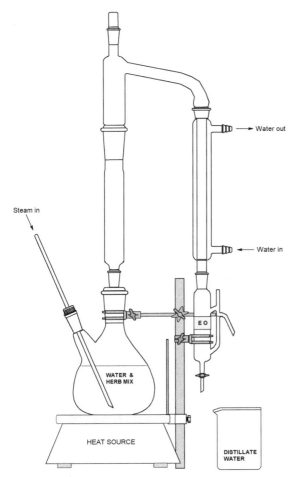

fig 38: the steam-injected hydro-distillation setup

The type of plant material best suited to this process tends to be tougher in nature (e.g. seeds/fruits, barks, roots etc).

Note: this is an advanced distillation technique requiring extra equipment to perform.

caution: working with pressurised steam can be very dangerous. This technique is not recommended for the beginner and must only be performed by those who are skilled and confident in working with this medium (and wearing the appropriate safety gear).

percolation (hydro-diffusion)

Although not technically a form of distillation, this process is similar in many respects to conventional distillation, as it uses steam to extract the aromatic essences from the plant material. However, in this case, it is dispersed through the plant material from the top of the plant chamber, rather than being introduced from the bottom (see fig 39: the percolation setup, below). In this way, rather than moving naturally upwards, it is forced down through the aromatic plant charge so that the steam can saturate the plant material more evenly; and gravity also helps in the extraction process. The resulting aromatic extracts contain some of the heavier, larger components that would otherwise be absent in the corresponding EOs (in a way, more like an absolute), and also smell much more like the original plant.

The water produced is more like that obtained via water infusion / decoction than a DW, and so can often be highly pigmented and can go off extremely quickly unless preserved. The resulting aromatic extract can then be separated from the water extract via the use of a sep funnel, before being washed with distilled water, dried (with magnesium / sodium sulphate) and then stored.

A benefit of this method is that the procedure time is shortened in comparison to other types of distillation. This means that for EOs that would normally require large amounts of time to process (such as wood and root oils), this is a much more economically-viable procedure.

Note: this is an advanced distillation technique requiring extra equipment to perform.

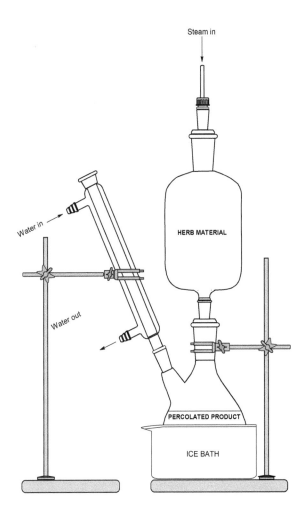

Steam in

Water in

Water out

HERB MATERIAL

PERCOLATED PRODUCT

ICE BATH

fig 39: the percolation setup

caution: working with pressurised steam can be very dangerous. This technique is not recommended for the beginner and must only be performed by those who are skilled and confident in working with this medium (and wearing the appropriate safety gear).

part 2:

skin-care products

definitions

macerated oils

Technically, a raw material, but can be also be used on its own as a skin-care product (the same is true of tinctures).

Macerated oils (MOs) are sometimes referred to as 'floral oils' (especially if the plant materials used are flowers or petals) or 'herbal oils', and are very similar to the infused oils (IOs - see page 79) mentioned in the first part of this book (i.e. vegetable oils (preferably refined) that have been infused with the aromatic and therapeutic compounds extracted from herbal material). The term macerate means 'to make soft by soaking / steeping in a liquid'.

The main difference between MOs and IOs is how they are produced. With IOs, the extraction into the vegetable oil medium is helped along with the addition of direct heat provided by a hotplate; with MOs however, only indirect heat from sunlight is applied to the process. Consequently, MOs usually take a lot longer to produce (weeks rather than hours). Even so, the absence of direct heat / high temperatures in the process enables the use of certain vegetable oils and the extraction of aromatic substances that may otherwise be susceptible to chemical change or degradation. In many cases the resulting products from these two processes can be used interchangeably, either on their own, or when incorporated into ointments, creams etc.

Note: see appendix 3: 'macerated / infused oil monographs' for more information).

tinctures

Tinctures (TRs) are alcoholic concentrated liquid extracts of the soluble constituents of herb materials obtained via the process of maceration (although you can sometimes find tinctures of other substances, such as iodine). Because alcohol is used as the solvent, water-soluble and most oil-soluble components are extracted and present with the TR. The process is similar to that of making MOs (see above) except that alcohol is used in place of vegetable oil, and the macerating material is best kept out of sunlight in a cool place. The alcohol used is ethanol (NOT methanol or isopropanol, as these are poisonous), and its action is not only to extract constituents from herb material but to also act as a preservative for the final product.

Various strengths of alcohol can be used to make herbal tinctures depending on the type of herbal material extracted. For example if you wanted to make a tincture from leaves, flowers or aerial parts in general, then a 25-37.5% alcohol is usually sufficient (e.g. vodka). However, if the herb material is of a high resin / oil content - for example myrrh or marigold - then anywhere up to 96% alcohol needs to be used to make sure the material is sufficiently extracted and that the constituents remain in solution.

Note: the purchase and use of 96% alcohol requires a government alcohol licence (see *resources* section).

The length of time the herb material is left to macerate may also depend on the type of material used (i.e. tougher material, e.g. barks or roots, is allowed to macerate longer in order to soften the material and gain optimum extraction).

Fresh herb material can be used to make tinctures but it is more common for dried material to be used.

The strength of herbal tinctures is usually given as a ratio; e.g. 1:5 – meaning that 5 parts alcohol was used to extract 1 part herb material (by weight).

Most tinctures can be used internally as medicine, although it is best to seek the advice of a qualified medical herbalist before doing so. Tinctures can also be used externally for a number of conditions or they can be incorporated into other products such as gels, balms, creams etc. However, you should only use small amounts owing to the drying action of the alcohol and its ability to separate out some products, such as creams (see appendix 4: 'herbal tincture monographs').

gels

Gels are water-based, semi-solid preparations, which can either be transparent, semi-opaque, or white. They are used for skin care (and more recently, hair care) applications, and are made by combining water with a 'thickening' agent (such as cellulose or xanthan gum – see appendix 6: 'recommended raw materials'), along with any EOs and moisturisers required to obtain the final product desired. As well as having the effect of thickening the water, these materials also have an astringent and protecting effect on the skin, without readily being absorbed.

Gels are able to hold a maximum EO / oil content of 5% and so are ideal for indications where fat- / oil-based preparations are not suitable (e.g. minor burns or sunburn).

Note: some people find that using gels regularly can start to dry out their skin. In these instances a small amount of vegetable oil / MO / IO can be added to the gel (1-3%) to reduce this effect.

ointments & balms

Ointments and balms are semi-solid skin preparations containing mainly oily ingredients (i.e. little or no water is added / present). They are made from combining an oil (such as vegetable oil or MOs / IOs) and a hard fat / wax (beeswax etc), along with any EOs / aromatic extracts that you wish to incorporate.

Balms (also called 'salves') are made the same way as ointments, the only difference being that balms (particularly lip balms) contain a higher proportion of hard fat / wax than ointments, which means that they are less likely to melt at higher temperatures and don't deposit too much oil onto the applied area.

The resulting products can be heavy and greasy in nature, which means that when they are applied to the skin they form a soothing, healing and protective layer that tends to stay on the surface of the skin longer than creams or lotions do.

Ointments and balms help to keep body heat and water in and so can be very effective when used to treat both areas of dry / rough skin (such as knees and elbows), as well as delicate areas of skin (such as lips and eyelids). Ointments and balms can also be valuable as a rub when treating rheumatism and dull aches that may occur in joints and muscles.

Note: ointments and balms should not be used to treat skin conditions that are hot, inflamed or weeping, as their application may make the condition worse!

creams & lotions

Simple creams and lotions are opaque, semi-solid skin preparations containing a stable mixture of oil (such as vegetable oil or MOs / IOs) and water (such as distilled water, DWs, herbal teas etc) via the use of emulsifiers. These preparations can be either 'oil in water' (where the proportion of water is greater than oil - producing the 'cosmetic' type creams available today), or 'water in oil' (where the proportion of oil is greater than water - producing thicker creams more commonly used in the past).

Other ingredients can also be added to adjust the feel and texture, colour, appearance, smell, and therapeutic action of the cream / lotion; and many creams / lotions also have preservatives added to them to extend their shelf lives.

Lotions are produced in the same way as creams except that they have a runnier consistency (they are therefore stored in bottles rather than jars), which allows them to be applied to larger areas of the skin, acting primarily as moisturisers. This is achieved by reducing the quantities of emulsifiers and oils present and increasing the quantity of water used to make the product.

Creams / lotions are easily absorbed by the skin and as a result they tend to feel a lot lighter than ointments / balms, and are more suited to application to the warm, damp areas of the body (e.g. groin). They also express a cooling / soothing action that makes them perfect for hot, inflamed, and weeping conditions.

making skin-care products

In this part of the book I refer to gels, ointments, balms, creams and lotions as base products. Aromatic extracts, EOs, DWs, TRs, MOs / IOs may either be used as they are but are often regarded as raw materials that may be added to the above base products to customise them for specific medical or aesthetic purposes.

In order make the base products detailed in this book, you must first gather together the equipment you will need to follow the procedures. You will find that a lot of the equipment needed can be found in the kitchen supply section of most local department / hardware stores. You can of course use any equipment you may already have handy (as long as it is cleaned / sterilised thoroughly). Personally, I prefer to have a separate set of utensils for making base products only, which I sterilise before each batch is prepared.

A lot of the raw materials needed to make these base products can be found at most herbal medicine and EO suppliers (see *resources* section). However, there are some ingredients that need to be sourced from specialist outlets (such as some of the emulsifier mixes). There are a few of these companies around in the UK, which sell both retail and wholesale amounts (see *resources* section).

equipment

The following is a list of equipment you will need to get started in making these base products.

scales and measures

When looking to buy these you will undoubtedly find that there is a wide selection of styles and types to choose from, with an equally varied price range. It is important to note here that they could possibly be your most important investment when it comes to making base products as they will enable you to recreate successful recipes and avoid unsuccessful methods of batch production. My preference is to go for a hard-wearing, laboratory-type digital scale (which allows operation from both batteries and mains supply – as batteries always seem to run out when you least expect and want it), that measures in grams (preferably accurate to the nearest 0.1g), and also has a 'tare' function (i.e. allows you to reset the scale to zero after the addition of each ingredient - so that everything can be measured into one container).

A measuring jug may be needed to accurately measure out any liquid ingredients (I use lab-style 100ml and 500ml Pyrex® measuring cylinders), and you may also need to buy a set of (preferably stainless steel) measuring spoons, or plastic measuring cups, for measuring out smaller amounts of ingredients.

hotplate

This can be similar to the one used in the distillation kit, and is needed to heat the ingredients before they are mixed together. I recommend that you get one that is electric (no open flames in the work area!), has a variable temperature control and preferably two plates / rings (better for cream / lotion making).

electric hand-held mixer

The next most important investment is a reliable mixer - mostly used for cream / lotion making, to help blend together the water and oil portions of the mix. Again, there are many types available on the market, but I recommend getting one with a strong motor (e.g. 600 Watt) and a detachable stainless steel shaft / blade, which makes cleaning and sterilisation a lot easier. If you do not have one of these you could always use a stainless steel fork or whisk, but the resulting product does not tend to be as smooth and creamy as if you had used a mixer.

water bath

Also known as a 'double boiler' or 'bain-marie'. This is the same as the water bath used in the reflux technique detailed in part 1 (page 79), and is needed to help gently heat / melt some of the more solid fat ingredients without them getting burnt or spoiled.

To do this you need to fill a stainless steel saucepan (not too full!) with water. You then place your mixing container (preferably stainless steel or Pyrex®) housing the raw ingredients (e.g. cocoa butter, beeswax etc), inside the saucepan, immersed in the water. The saucepan is then gently heated on a hotplate until the ingredients in the inner container have melted. In most cases you will only need one of these, but when you start experimenting with your own recipes you may wish to have two handy.

thermometer

This is essential for monitoring the temperature of your products as they are being made. As you will see in the following sections, it is very important to make sure that the ingredients have reached the right temperature before they are mixed together (more so for creams). Failure to correctly monitor the temperature may result in the separation and subsequent spoiling of your product. To start with, you can get away with just using one thermometer (measures up to 110°C), but you may choose to get another one as you progress to making your own recipes.

Note: don't be tempted to use the thermometer as a stirrer, as it may break and ruin the product.

mixing containers

These are what the ingredients will be heated and mixed in. I prefer to use laboratory-style, heavy-duty Pyrex® measuring beakers (thick walled) of 1000ml and 2000ml capacity, but conventional kitchen-style Pyrex® or stainless steel bowls / measuring jugs (of appropriate size for your product batch) could be used instead (available at most department / hardware stores). You will need to get two of these for cream / lotion making - one to heat the oil phase and one to heat the water phase simultaneously.

stirrers and spatulas

The stirrers should be sterilisable to reduce the possibility of contamination, and preferably made of either glass or high temperature / chemical resistant polypropylene; they are useful for stirring the ingredients as they are heated / melted before they are mixed together.

Spatulas are very useful for helping to scoop out any product that may be left clinging to the sides of the mixing containers after pouring it out into your storage containers. You can get different types of spatulas, but I prefer the hard rubber type as these are flexible enough to effectively scoop around the mixing containers I have without scratching or chipping the glass.

You should never use your finger to remove creams from their storage containers as this can introduce fungi or bacteria and reduce their shelf life considerably. It is therefore a good idea to get some small wooden spatulas (single use), and use these instead to remove creams from their storage containers for use or testing.

storage containers

These are what you are going to put your final product in once it has been made. As you can imagine, there are a variety of containers that you can use - ranging from plain glass or plastic jars to fancy hand-blown dispensers. As long as the container has been sufficiently cleaned / sterilised before use it should be OK to use. I like to use amber-tinted glass jars (plastic can be permeable and can sometimes affect the quality of the product) with corrosive-resistant, airtight lids for my creams. For lotions I use tinted glass containers with plastic, applicator-type lids. For more information on this subject see page 71: 'storage and shelf-life'.

heat-proof mat

Useful for putting hot (flat-bottomed) glassware on when it needs to cool down (and also for protecting your work surface). These can be purchased from lab supply stores in various sizes, and asbestos-free. However, you can quite easily use a normal ceramic tile (for bathrooms and kitchens) instead.

raw materials

The following is a list of raw materials you will need to get started in making these base products. Some of these materials can be obtained or made from either animal or plant sources, or are synthetically (or semi-synthetically) produced. Wherever possible I prefer to use materials that are derived from plant sources only, with the exception of beeswax.

Note: see also appendix 6: 'recommended raw materials', and appendix 7: 'raw materials to avoid'.

fats

These materials are present in two main forms - liquid fats and solid fats. Both of these types of fats help to give your base products rich moisturising capabilities, and also allow other fats / fat-soluble materials (such as aromatic extracts / EOs) to be absorbed / incorporated into your final product (some may also help with the absorption of these added materials through the skin).

liquid fats

These types of fats are very common and the 'raw' and cold-pressed oils are extensively used for culinary purposes.

vegetable oils

Vegetable oils are extracted from plant sources (i.e. from nuts, seeds, kernels etc) via pressing (preferably cold pressed). When using these to make base products it is often recommended that you use the 'refined' type of vegetable oils as they have been further processed to remove the product's odour (and possibly colour), which may otherwise compete with any other aromatic materials you may add, and therefore taint your final product. However, some may argue that this further processing may also remove or denature many of the naturally-occurring therapeutic compounds present in the 'raw' oils. It is also important to note that the more 'raw' an oil is, the shorter the shelf-life will be. These materials can also help to add fluidity to your product without making it too watery.

macerated / infused oils

These have already been discussed (see page 89). They perform a similar job to vegetable oils, except that they also have the added therapeutic benefits of the herbs with which they have been infused / macerated.

Note: also see appendix 3: 'macerated / infused oil monographs'.

solid fats

These types of fats can be extracted from plant (e.g. cocoa butter), animal (e.g. beeswax) or mineral (e.g. paraffin) sources, or can be made synthetically. The addition of these materials helps to thicken up the base product (some may also act as mild emulsifiers) so that it becomes more 'spreadable' and less runny, as well as adding their own therapeutic properties (e.g. moisturising) to the final product.

waters

These help to (re)hydrate the skin as well as allowing any water-soluble compounds (such as water-soluble volatiles) to be present in the final product (e.g. gels, creams / lotions). They may also help these compounds to be absorbed into the skin, and are useful for diluting a product that may otherwise be too greasy / heavy to use for the application desired.

spring / mineral / filtered water

Spring and mineral waters are commonly found bottled in any supermarket, and filtered water can be made at home with tap water and a variety of water filters. A lot of people prefer to use this type of water as they believe that plain distilled water (see below) is lifeless.

Note: these waters should be heated above 85°C for at least 10 minutes before use. If not, then a higher percentage of preservatives need to be added to the gel / cream / lotion - otherwise it will not keep for more than a week or two.

distilled water

This is plain water that has gone through the process of distillation (hydro-distillation) to remove any minerals / salts that may otherwise impede its solubility when applied to a product.

The process of distillation also helps to sterilise the water, but if it is not used immediately after being distilled it may need to be re-sterilised (but maybe for not as long as for spring / mineral / filtered water).

distillate waters

These are explained in more detail in part one of this book (page 20), and in appendix 2: 'distillate water monographs'); they perform a similar job to distilled water except that they also have the added aroma and therapeutic benefits of the herbs they have been distilled from.

As with the distilled water, these may also need to be re-sterilised before use, if not used directly after being distilled.

decocted / infused / macerated / percolated waters

These are basically herbal 'teas' made either with the application of heat (e.g. decoction, infusion, and percolation) or without (e.g. cold maceration). These do a similar job to the DWs discussed above, except that they may be coloured and contain more (larger / heavier) compounds extracted from the herb material that may otherwise be absent via the process of distillation.

Note: if these are used then a higher percentage of preservatives always need to be added (usually at double their regular dose), otherwise the products will spoil extremely quickly.

emulsifiers

See also appendix 5: the chemistry of emulsification.

Of all the raw materials used in the manufacture of base products, these are the ones that are most shrouded in mystery. Emulsifiers are essential to the production of base creams / lotions, because they enable the otherwise 'non – mixable' fat and water phases present in the mix to blend together and stay mixed together over a long period of time (hopefully without separating again!). Because of their action they can have a great impact on how the end product may feel when applied to the skin, depending upon what emulsifiers (or combinations of emulsifiers) are used. As a result, a huge amount of money is spent by corporations on employing chemists to discover and patent the right combinations of emulsifiers to make creams feel just right (passing these costs on to the purchaser of course). Naturally they can be a little reluctant to give away the secrets of their exact recipes.

Emulsifiers can come from natural sources (both plant and animal), but recently more and more are being produced synthetically (or semi-synthetically). There are two main types of emulsifiers; ones that help oil to mix with water, and ones that help water to mix with oil (see below).

Most modern base cream / lotion formulations contain both types of emulsifiers (one helps to regulate the action of the other), and their ratios are adjusted according to the other ingredients present within the mix, and to deliver the properties required in the end product.

oil-in-water emulsifiers

These types of emulsifiers are most commonly seen in 'modern' cream / lotion formulations, where there is at least 60% water or more (i.e. 40% oil or less) present in a mix. An example of this type of emulsifier is sodium stearoyl lactylate (also known as sodium stearoyl-2-lactylate); which is safe to use, edible, and can be semi-synthesised from coconut and palm oil). They are added to the water phase of the base cream- / lotion-making procedure.

water-in-oil emulsifiers

These types of emulsifiers are used to produce the thicker / oilier type cream / lotion formulations used more in the past, where there is 50-60% oil or more (i.e. 40% water or less) present in the mix. Examples of this type of emulsifier are glyceryl stearate or glyceryl monostearate; which is also safe to use, edible, and can be semi-synthesised from coconut and palm oil. They are added to the oil phase of the base cream- / lotion-making procedure.

Note: as with the preservatives below, there are some types of emulsifiers that can be harmful or cause allergic reactions in some people and should therefore be avoided (e.g. borax (sodium borate), and triethanolamine (TEA)). See appendix 7: 'raw materials to avoid' for more details.

preservatives

Preservatives causes a lot of headaches for commercial producers of base products (especially gels and cream / lotions), because no matter how careful you are with cleanliness and sterilising during manufacture (although this is always advised!), your products will always be susceptible to some level of bacterial and / or fungal infestation. In order for these organisms to survive and thrive, they need water. Therefore, the more water / moisture present in a product, the more likely it is to become contaminated with these organisms and so more preservatives are needed to counteract this. Most commercial producers therefore choose to use preservative mixes in their products to extend their self life. Unfortunately, a lot of the common types of preservatives applied may cause illness / allergic reactions in some people who use them (e.g. kathon CG, and 2-bromo-2-nitropropane-1,3-diol). See appendix 7: 'raw materials to avoid' for more details.

There are some alternatives available to these commonly-used preservatives. For example, you can use single or combinations of anti-microbial EOs / DWs - for example tea tree EO (using a minimum of 2% in the product) or pure rose geranium DW as the water phase. Alternatively you could increase the acidity of the product to below pH4 (be warned: adding too much of an acid can make creams separate). By doing this you can sometimes help creams to last for up to 6 months with minimal use of preservatives (depending of course on what else is in the product).

Even so, these techniques may only be effective if the products are made under strict sterile conditions, and sometimes they may be impractical to incorporate into the mixes (e.g. you can't really use tea tree EO if you are making a rose cream). Also, they won't be effective if any water-based herbal extracts (such as decocted / infused waters etc) or other highly-perishable ingredients are added to the mix.

In order to be able to limit the use of preservatives in your products, I recommend keeping your work area and equipment meticulously clean, following GMP (good manufacture practice - see *resources*) standards or possibly different types of storage containers that may help to give protection from contamination (see page 141: 'storage and shelf-life of skin care products' for more information).

If you do have to use a preservative (such as when you are using decocted / infused waters in your products), although not ideal, I would recommend using either sodium benzoate, or preservative mixtures such as euxyl K700 (Preservative K) to give a broad spectrum anti-microbial effect, and therefore extending your products' shelf life to 1½-3 years. Even so, you should be aware that some people can still have allergic reactions to these products.

For more information about each of these preservatives and how they can be applied, please refer to appendix 6: 'recommended raw materials'.

other ingredients

Once you have mastered making your own base products you can start to add other ingredients to personalise your end product or adapt its aesthetic or therapeutic effects to meet your needs.

essential oils

Production of these is covered in detail part one of this book. They can be applied either on their own or in combination to impart a particular aroma to a base product. When used and blended professionally they can also be very powerful therapeutic agents.

aromatic / herbal extracts

Again, these are covered in more detail in part 1 of this book. As with EOs, these materials can be used to add aesthetic and / or therapeutic properties to your base products.

resins / gums

Examples are frankincense and myrrh (these may need to be powdered or mixed with agents that help keep them in a liquefied state- or used as TRs).

expressed and percolated oils

These are similar to the EOs; however, care should be taken as they often contain compounds that are not present in distilled products (e.g. the compound bergapten which is present in bergamot oil can cause phototoxicity).

absolutes, tinctures, CO_2 extracts, and florasols

Again, as with the expressed and percolated oils, a greater proportion of compounds may be extracted from the plant material, some of which may cause sensitivity in some people. These materials also tend to be very concentrated, and so only small amounts need to be used to obtain the effects desired. They may also colour the product - sometimes undesirably.

vitamins and minerals

These are often added to base products for their therapeutic properties on the skin (an example of a commonly used vitamin is tocopherol (vitamin E) and its derivatives which are powerful anti-oxidants and help to extend the shelf life of oils).

texture enhancers and moisturisers

These materials are sometimes added to base products (again, more often to creams / lotions) to improve the feel and flow of the product on the skin and to enhance their moisturising properties. For example, compounds such as 'natural moisturising factors (NMF)' (a mixture of deep moisturising compounds – see appendix 6: recommended raw materials), and glycerine (providing moisture to the skin surface) are commonly used to improve the moisturising / (re)hydrating properties of a cream / lotion. There are also other compounds (such as xanthan or cellulose gum) that are used to produce gels and to thicken creams / lotions.

botanical colourants

The application of certain botanical extracts (especially decocted / infused waters etc) in your mixes can result in a product that exhibits an undesirable colour (commonly a grey / green hue). In order to counteract this you can add small amounts of other herbal extracts to obtain an agreeable colour (see page 138, problems with the colour of products).

further laboratory health & safety

As with the procedures detailed in part one of this book, all workers must assume a responsible attitude to their work. They must also try to avoid any careless or rushed behaviour which may lead to an accident and possible injury either to themselves or others. They should always pay due consideration to what is going on around them and be aware of any possible dangers arising from their procedures.

The same health and safety precautions as for part one should be applied here, along with the added precaution below.

know your raw materials

I would encourage you to experiment with your own versions of the formulations set out in this book in order to meet your own particular needs. However, before you do so it is vital that you know as much as you can about all of the raw materials with which you are working. This is not only important in terms of how you apply a raw material into a product (a lot of money can be wasted on incorrect / inappropriate application of raw materials which may result in spoiled products), but also what the final product is going to be used for (e.g. you wouldn't want to add EOs to a product that may have contraindications to the skin type you have designed the end product for, either outright or at the particular dosage used).

You may also want to consider where your raw materials are sourced from (i.e. animal, botanical, synthetic), as you may not want to unwittingly incorporate a material into your mix that is ethically / ecologically undesirable.

appendix 6: 'recommended raw materials', and appendix 7: 'raw materials to avoid' may be of help to you now, but it is advisable to keep up to date with any information about the raw materials you use, as this is constantly changing.

Note: I recommend that you try to obtain 'Material Safety Data Sheets' (MSDSs) for all the raw materials you buy. These sheets should be given freely by your supplier and contain valuable detailed information about each material, including its purity, source, storage conditions, shelf life and any health considerations.

how to make
skin-care products

macerated oils

This procedure is easy to carry out. You will need the following equipment: scales; 2 dry, sterilised one-litre wide-mouth glass containers with airtight sealed lids (jam / preserving / pickling jars are perfect for this); some clean muslin cloth (a wine / fruit press is better but not essential); a funnel; some filter paper; and a dry, sterilised tinted glass storage bottle with lid.

You will also require the following raw materials: 600g of a vegetable oil or mix of vegetable oils (preferably refined), and 300g of herb material (partially dried or dried, and coarsely chopped - although some flowers / petals may be used whole).

step 1

Fill one of the one-litre wide-mouth glass containers (adequately cleaned / sterilised / dried beforehand) with 300g herb material (don't pack too tightly), followed by 600g vegetable oil(s), making sure that the herb material is submerged completely (see fig 40: step 1).

Note: You may want to gently tap the container on your work surface to try to dislodge any air bubbles that may be present.

step 2

Seal the container (see fig 41: step 2a) and leave it in direct sunshine for up to 8 weeks (or until the herb material starts to turn brown) - see fig 42: step 2b.

Note: You should be aware that when the oil is warmed it will expand, causing a possible overflow if the jar is too full. It is therefore recommended that it be placed on a plate or saucer to collect any overflow. It will also greatly improve the extraction if you shake the jar vigorously at least once a day.

step 3

Strain the herb / oil(s) mix through a muslin cloth (remember to give it a good squeeze, or alternatively you can use a fruit / wine press); saving the oil in the other one-litre wide-mouth glass container (also adequately cleaned / sterilised / dried beforehand), containing another 300g of the same herb material (see fig 43: step 3). The old / spent herb material (known as the marc) can now be simply thrown away.

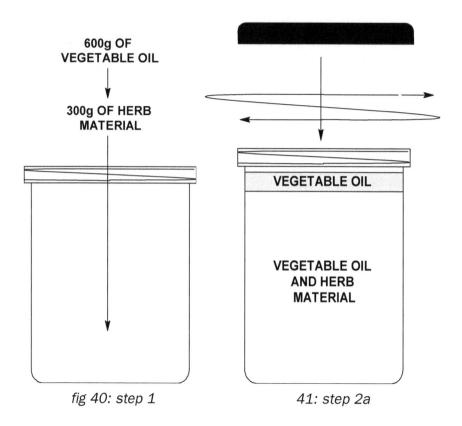

600g OF
VEGETABLE OIL

300g OF HERB
MATERIAL

VEGETABLE OIL

VEGETABLE OIL
AND HERB
MATERIAL

fig 40: step 1   *41: step 2a*

step 4

Repeat Steps 2-3 as many times as is needed (usually 2-4 times) to obtain the strength of product desired (as in the reflux process when making IOs – see page 79).

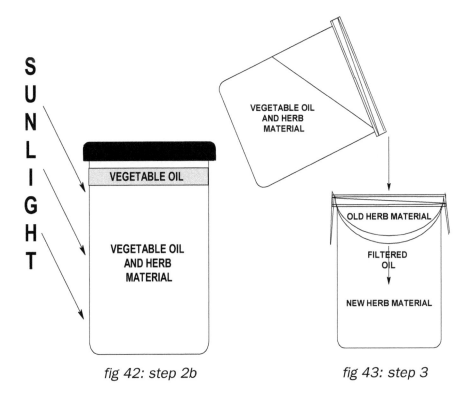

S
U
N
L
I
G
H
T

VEGETABLE OIL

VEGETABLE OIL
AND HERB
MATERIAL

VEGETABLE OIL
AND HERB
MATERIAL

OLD HERB MATERIAL

FILTERED
OIL

NEW HERB MATERIAL

fig 42: step 2b   fig 43: step 3

step 5

After final straining it is preferable to filter the macerated oil further through filter paper in a funnel in order to remove sediment, improve the clarity of the final product and extend its shelf life. Due to the viscosity of the MO this may take several hours and several changes of filter paper.

Remember to properly label your MO (parts and state of herb material used – dried / partially dried; oil(s) used; concentration; strength; date of manufacture; expiry date; batch no. etc.). The volume of finished product will be approximately 470ml and have a concentration of 1:2 (i.e. 1 part herb to 2 parts oil(s) by weight), and a strength of X times (written 2x, 3x, etc, depending on the no. of times steps 2-3 were repeated). It should be stored in a tinted / airtight glass bottle and can either be used as is or incorporated into one of the other products mentioned in this book.

caution: if using herb material grown or wild-crafted by you then it is your responsibility to make sure that it is accurately identified, harvested at the correct time and that the correct part of the plant is used, and in the right concentrations - especially if the products you make are going to be given / sold to friends, family or members of the public.

Notes:
Any drying or partial drying of herb material should be conducted under a low heat (i.e in a dark room at 24-26°C for 1-2 days) to avoid losing many of the volatile compounds.

Powdered herbs can be used in this process, but you should be aware that the final filtration may be a slow and lengthy process.
Most MOs will keep for about a year but you can add vitamin E oil (0.5-1%) to the oil before macerating the herb material to help extend shelf-life (make sure it is mixed in well).

As with IOs, MOs can also be made using fresh herb material. However, due to higher levels of moisture present, less herb material (approx ½ as much) can be processed in the same amount of oil(s), which in turn leads to weaker extracts.

The higher levels of moisture present during maceration may also lead to increased potential for yeast / mould growth, and as with IOs, you may find that a watery green liquid is present at the bottom of the oil at the end of the process. This liquid must be thoroughly separated out and thrown away; otherwise it will go off and quickly spoil the MO / IO.

It may be advantageous to calculate the water content of the herb material used (see section on making TRs below for more information on how to do this), and then experiment with different levels of partial drying (i.e. dry herb material by 25%, 50% etc of its water content) to see how this effects the quality of the end product. In this way a good balance between drying the herb material just enough to avoid the above problems with excess moisture and drying out too much and losing yield on some of the essential oils and other volatile compounds that you want to extract. For example, marigold flowers are best macerated when they have been dried to 50% of their water content (usually 2-3 days at 24-28°C, or until the petals start to curl). Chamomile flowers are also best macerated at 50% of their water content (although the drying time should be shortened to reduce the loss of essential oil yield). St John's Wort flowers on the other hand are best used after just one day of drying.

Notes:

When some types of herb material are processed this way they can cause the vegetable oil used to change colour. This is fine, and is most notable when making damask rose petal MO (you get a pink colour), and St John's Wort MO (you get a blood-red colour).

It is recommended that rounded shoulder jars (with 110mm diameter screw top caps) be used when making larger volumes of MOs in order to facilitate the 'compacting' of the herb material and therefore help to hold it at or below the surface level of the oil(s) used.

Vegetable butters may also be used in this process, although you should be aware that these need to be incubated at a higher temperature during the maceration process in order to keep the butters in a mobile liquid state for the extraction to be effective. Many different vegetable oils (and even mixes of oils) can be used to make MOs. In order to choose which one(s) you want to use it may be useful to consider the following:

1. the nature of the herb material constituents you want to extract (i.e. is the herb material high in essential oils, or resins etc?); it is best to choose oils that are most like those present in the herb material in order to get the best possible extraction.
2. what you are going to use the MO for (i.e. are you going to use it as it is or incorporate it into another product); the suitability of the oil(s) chosen needs to be taken into account.
3. the beneficial effects of the oil(s) in their own right (i.e. moisturising / nourishing / healing properties).
4. the chemical stability against oxidation of the oil(s). Generally, the more unsaturated an oil is - i.e. contains higher levels of duo-unsaturated (Omega 6) and poly-unsaturated (Omega 3) fatty acids - the more likely it is to oxidise and go rancid, and so the shorter the shelf life. This may be counteracted to a certain extent by adding a high % of vitamin E oil (see above).
5. any allergic considerations (such as oils derived from nuts - e.g. almond oil).
6. the age of the oil; there is no use buying oil(s) that are already a year old as this will mean that the shelf-life of your product will be greatly reduced. Make sure you buy from reputable companies and ask exactly how old the oil(s) are (from date of pressing), and how they have been stored.
7. The availability and cost of the oil(s).

It is recommended that you refer to the following books: *Liquid Sunshine* by Jan Kusmirek and *Fats that Heal, Fats that Kill* by Udo Erasmus (see *resources*) for more information on the subject.

herbal tinctures

This procedure is very similar to making MOs. You will need the following equipment: scales; a measuring cylinder; a sterilised, suitably-sized wide-mouth glass container with an airtight sealed lid (jam / preserving / pickling jars are perfect for this); some clean muslin cloth (a wine / fruit press is better but not essential); a funnel; some filter paper; a glass beaker; and a sterilised tinted glass storage bottle. You will also require the following raw materials: 1000ml (approx. 920g) of alcohol – e.g. ethanol (vodka - 37.5% - is used in this example) and 170g of dried herb material (chopped / powdered).

step 1

Using the scales, weigh out 170g of the coarsely chopped (or powdered) dried herb material and put into the wide mouth jar (don't pack too tightly). Then use the measuring cylinder to measure 1000ml of vodka (37.5%) and pour into the wide mouth jar on top of the herb material.

Note: make sure that the container has been adequately cleaned / sterilised beforehand.

step 2-3

Seal the container (see fig 45: step 2) and shake the jar vigorously for 1 minute to ensure the contents are thoroughly mixed. Store in a cool dark place for 10-21 days, or 30 days if the plant material is especially fibrous / resinous (see fig 46: step 3).

Note: in some cases the herb material may need to be squashed down with the back of a sterilised spoon to ensure that it is submerged beneath the level of the alcohol. It will also greatly improve the extraction process if you shake the jar vigorously at least once a day for 1-2 minutes.

step 4

Strain / squeeze as much of the herb / alcohol mix as you can through the muslin cloth (alternatively you can use a fruit / wine press), saving the liquid alcohol extract (the menstruum) in the glass beaker (see fig 47: step 4) and throwing the old / spent herb material (the marc) away.

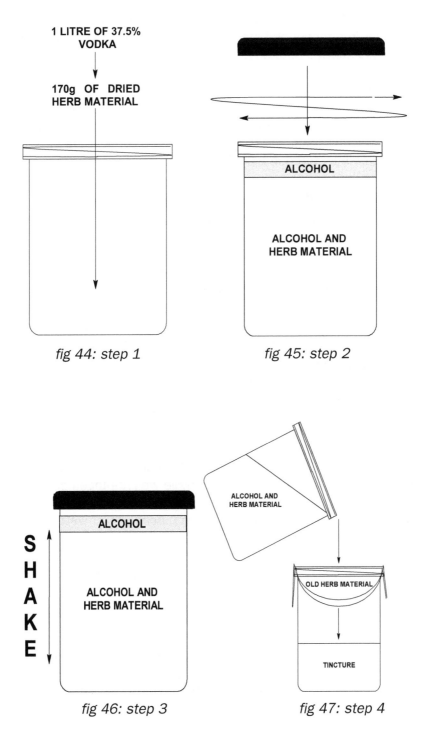

1 LITRE OF 37.5% VODKA

170g OF DRIED HERB MATERIAL

fig 44: step 1

ALCOHOL

ALCOHOL AND HERB MATERIAL

fig 45: step 2

S
H
A
K
E

ALCOHOL

ALCOHOL AND HERB MATERIAL

fig 46: step 3

ALCOHOL AND HERB MATERIAL

OLD HERB MATERIAL

TINCTURE

fig 47: step 4

You may now filter the liquid extract further by passing it through some filter paper in a funnel and storing the final extract (TR) in the sterilised tinted glass storage bottle.

Note: remember to properly label your TR. The finished product will be approximately 950ml and have a concentration of 1:5 (i.e. 1 part herb to 5 parts alcohol by weight; also see section on tinctures on page 89) in 37.5% alcohol, and will keep for 2-3 years (depending on the herb material used) if stored in a cool, dark place.

a note on making TRs from fresh herb material

The procedure for making TRs from fresh herb material is essentially the same as for making them from dried herb material. The main difference is that in order to get a more accurate concentration and alcohol content in your final product you need to take into consideration the amount of water present in the fresh material. In order to do this you need to weigh out 3 small samples of the fresh herb material you are going to use (it is better to weigh out the same amount for each sample to make the maths easier). Note this weight down, coarsely chop, and then dry it out as much as you can (either naturally in the sun, in an oven set on low heat, or in a food dehydrator).

Note: when the small twigs or leaves in the sample break sharply, it's a good sign that the sample is dry enough.

When the samples are thoroughly dried they are then re-weighed and an average is taken (i.e. add all three sample weights together and divide by 3). This value is then divided by the original weight and multiplied by 100 to give a (rough) estimate as to the percentage of water present in your fresh herb material.

For example if the dried samples weighed 2.5g, 3g, and 3.5g, then the average dry weight would be:

$$2.5 + 3 + 3.5 = 9 \qquad\qquad 9 \div 3 = 3$$

Therefore the average dry weight for your herb material is 3g.

If the fresh weight of each sample weighed 6g the calculation would be:

$$3 \div 6 = 0.5 \qquad\qquad 0.5 \times 100 = 50\%$$

Therefore your fresh herb material has a water content of (roughly) 50%.

If 184g of the same fresh herb material was used, 50% of it would be water (i.e. 92g of water).

As 1g of water equals approx. 1ml (0.9982g to be precise), this would translate as approx. 92ml of water (although if you were working with larger quantities it may be better to use the more precise value to avoid insufficient dilution of the alcohol).

Note: remember that this is still a rough estimate of the water content of the herb material, as it is not just water that evaporates off during the drying process - essential oils and other volatiles do too. However, this is the best method we have for approximating water content.

In order to make up one litre of 30% alcohol from 96% alcohol (ethanol) you would have to mix together 312.5ml of 96% alcohol with 687.5ml of water (preferably distilled and definitely sterilised).

As there is already 92ml of water present in the fresh herb material this would mean that only an extra 595.5ml of water would be needed, along with the 312.5ml of 96% alcohol to produce a tincture that has a concentration of 1:5 in 30% alcohol.

This is all very well, but not many people can easily get hold of 96% alcohol unless they have an alcohol licence (see *resources*). However, this same principle can be applied to using 37.5% vodka by using the chart below (see appendix 8 for tincture ratio chart using 96% ethanol).

tincture ratio chart using 37.5% vodka		
water required (ml)	alcohol required (ml)	final strength of TR (%)
333.3	666.7	25
200	800	30
66.7	933.3	35

Therefore, in order for you to make 1 litre of 1:5 tincture (30% alcohol), all you need to do is add 108ml of water and 800ml of 37.5% vodka to your fresh herb material.

Note: it is important that the alcohol and water are thoroughly mixed together before adding to the herb material because they tend to 'layer' very easily when combined. It is also important to note that the higher the % alcohol of ethanol used, the less 1 litre of it weighs (this is because alcohol is less dense than water). I have therefore included the weights of 1 litre of ethanol at commonly used alcohol percentages for TRs so that you may adjust the amount of herb used (either dried or predicted dried weight if using fresh herb material) in order to keep the ratios correct (see chart below).

Note: in order to get an idea of what strength to use and at what concentration for a particular herb, you can look in herbal medicine supply catalogues (see resource section suppliers) or refer to herbal pharmacopoeias. Generally, the higher the fat / oil / resin / wax content of the herb material, the higher the % of alcohol used to ensure the extracts remain in solution.

| weight of 1 litre of ethanol at varying % (at 20°C) ||
% alcohol	weight of 1 litre (g)
100%	789.0g
96%	797.4g
95%	799.5g
90%	809.9g
60%	872.7g
50%	893.6g
45%	904.1g
40%	914.5g
37.5%	919.7g
37%	920.8g
35%	925.0g
30%	935.4g
25%	945.9g

Please note that these values can change considerably with varying degrees of environmental temperature.

So for example, if you wanted to make 1 litre of 1:5 tincture at 96% alcohol (i.e. weighing 789g) you would use 157.8g of dried herb material. However, if you wanted to make a 1 litre of 1:5 tincture at 25% alcohol (i.e. weighing 945.9g) then you would use 189.18g of dried herb material.

Note: generally, you should aim to prepare a stronger tincture when using fresh herb material (i.e. 1:2 or 1:3), as more of the weight is water and therefore may produce a slightly weaker extract. Also, a higher % of alcohol is usually desired (at least 30%) to reduce the likelihood of spoilage.

The calculation for the percentage of water should also be done for every separate sample / batch used (even if it is the same species), as water content may differ from year to year depending on the weather conditions, how / where it is grown and how / when it is harvested etc.

You should also be aware that when harvesting fresh herb material, it starts to lose its water content immediately (especially after it is coarsely chopped, when it can lose anything up to 6% in weight in one hour!). Some herb material may also be susceptible to rapid oxidation upon exposure to air (such as chopped elecampane root). It is therefore imperative that once harvested, the herb material is processed and added to the alcohol as soon as possible.

caution: TRs should never be made with less than 25% alcohol, as this may severely impair the alcohol's ability to act as a preservative. It is also important that any water added to adjust the alcohol % is sterilised by boiling for at least 10 mins before use, otherwise the TR may become contaminated and unsafe to use.

Notes:
100% ethanol is very rare to obtain by the process of distillation alone (the highest % that may be obtained via distillation is 96%). This is because of ethanol's high affinity for water, which makes it very reluctant to give up its last 4% of water. In order to get 100% ethanol it must undergo a series of chemical drying processes, which makes it very expensive to buy. It is also very difficult to work with, because as soon as it is exposed to the open air it will start absorbing moisture from the atmosphere, turning it back to 96% again.

Other types of menstruum may be used in place of the alcohol / oil. For example vinegars, glycerine, and honey may be used to create different types of extracts (only use dried herb material with honey to avoid the excess waters causing fermentation).

Other types of 'beverage' alcohol can be used to make TRs (such as brandy, whisky, rum etc.), although you must bear in mind that because the alcohol present is already quite saturated with flavours, fragrances and colours, their ability to extract and hold further herbal constituents is limited, which in turn may result in the production of a weaker TR.

Any drying of bulk herb material should be conducted under a low heat (i.e in a dark room at 24-26°C for 1-2 days) to try to prevent the loss of too many of the volatile compounds.

It is recommended that rounded shoulder jars (with 60-70mm diameter screw-top caps) be used when making larger volumes of TRs in order to facilitate the 'compacting' of the herb material, which helps to hold it at or below the surface of the alcohol.

caution: if using herb material grown or wild-crafted by you, then it is your responsibility to make sure that it is accurately identified, harvested at the correct time and that the correct part of the plant is used, and in the right concentrations - especially if the products you make are then going to be given / sold to friends / family or members of the public.

caution: TRs can be very strong medicine, and so the above procedure has been given in order to understand the process of TR making for external application only. Some TRs can be very potent or even dangerous if taken at the wrong dosage. If you do want to take TRs internally then please consult your local qualified medical herbalist (contact the NIMH in the resource list) for guidance.

TRs can either be used alone, or incorporated into other products (see appendix 4, herbal tincture monographs for more details).

gels

This is the simplest base product to make. To follow this procedure you will need the following equipment: scales; a measuring cylinder; a mixing container; a polypropylene / glass stirrer; an electric hotplate; a heatproof mat / ceramic tile; a thermometer; an electric hand-held mixer (or a whisk); a spatula; and storage containers. You will also require the following raw materials: xanthan gum; a water of your choice (in this example distilled water is used); glycerine; and any other raw materials you wish to add (e.g. EOs, vegetable oil / MOs, preservatives etc).

caution: this procedure requires the application of heat and the manipulation of hot liquids. You should wear suitable eye and hand protection to avoid possible injury from burns.

Note: it will help you a lot if you can to get your sterile storage containers ready before you start this process.

step 1

Using the scales, weigh out 5g of xanthan gum (slightly more can be used to create a thicker gel if required) and set aside. Then use the measuring cylinder to measure out 470ml of distilled water, and pour this into the mixing container along with 10ml of glycerine (see fig 48: step 1).

Also weigh / measure any of your other ingredients out now.

step 2

Set up your hotplate on a heatproof work surface, and place the mixing container from step 1 on top (remember to use the wire screen ceramic heat dissipater - ceramic side up). Turn on your hotplate, and with gentle stirring, heat the ingredients until the temperature of the water / glycerine mix reaches 36°C (use the thermometer to accurately measure this). See fig 49: step 2.

step 3

When the water / glycerine mix reaches the desired temperature, remove from the hotplate using heatproof gloves, set aside on a heatproof mat / ceramic tile, and turn the hotplate off. Then gently sprinkle the xanthan gum, a little at a time, into the water / glycerine mix whilst using the mixer (on lowest speed) / whisk to blend the materials together (see fig 50: step 3).

Note: it is important that you do not mix too vigorously or for too long, or introduce too many air bubbles into the mix, as this will result in a product that is white and has the consistency of chewing gum! To prevent this you should try to touch the bottom of the mixing container with your electric hand-held mixer / whisk, and move in a side-to-side motion, rather than up and down.

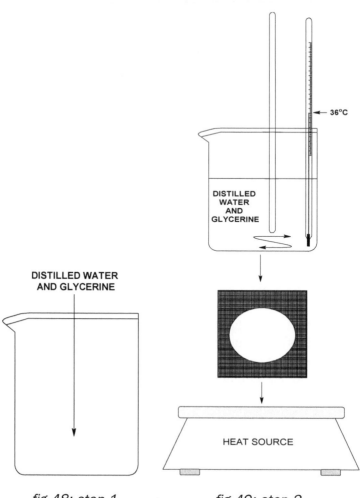

36°C

DISTILLED
WATER
AND
GLYCERINE

DISTILLED WATER
AND GLYCERINE

HEAT SOURCE

fig 48: step 1   fig 49: step 2

step 4

Whilst still blending, you can now add and stir in the remaining ingredients to your base gel (see fig 51: step 4), and then pour into your storage containers (formulation makes approximately 500ml of product). Remember to make sure the temperature is 25°C or below before adding EOs, to avoid evaporation.

Note: if you add strong anti-microbial EOs to the gel (e.g. tea tree EO) in a concentration of 4-5%, then it is usually not necessary to add any preservatives (depending on what else is added to the mix).

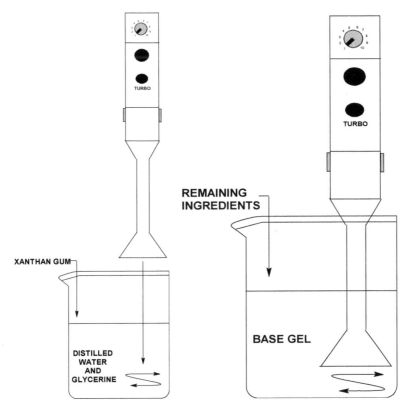

<table>
<tr><td>XANTHAN GUM</td><td></td></tr>
</table>

REMAINING
INGREDIENTS

XANTHAN GUM

DISTILLED
WATER
AND
GLYCERINE

BASE GEL

fig 50: step 3   fig 51: step 4

ointments

For this procedure you will need the following equipment: scales; a measuring cylinder; a mixing container; a water bath (i.e. a stainless steel saucepan with water in it); an electric hotplate; heatproof mat / ceramic tile; a polyproplylene / glass stirrer; an electric hand-held mixer (or a whisk); a spatula; a thermometer; and storage containers. You will also require the following raw materials: beeswax; veg oil(s) of your choice (e.g. sweet almond oil); and any other raw materials you wish to add (e.g. EOs etc).

caution: this procedure requires the application of heat and the manipulation of hot wax / oil. You should therefore wear suitable eye and hand protection to avoid possible injury from burns.

Note: it will help you a lot if you can to get your sterile storage containers ready before you start this process.

step 1

Using the scales and measuring cylinder, weigh out 80g of beeswax, measure out 420ml of vegetable oil(s), and put these together into your mixing container (see fig 52: step 1). Also weigh / measure any of your other ingredients out now.

step 2

On a heatproof work surface, set up your water bath on your hotplate; make sure that there is sufficient water present in your saucepan (not too much though!), and occasionally add more water when needed to avoid it boiling dry. Put the mixing container holding your ingredients from step 1 into the water bath. Turn on your hotplate, and, whilst stirring, heat the ingredients until the beeswax melts into the vegetable oil (see fig 53: step 2).

Tip: use coins to lift the beaker off the bottom of the saucepan to avoid bumping when heating.

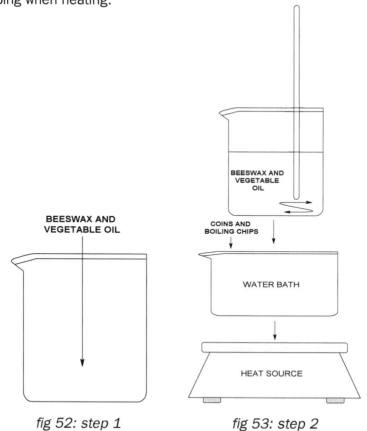

fig 52: step 1   fig 53: step 2

step 3

Carefully remove the mixing container from your water bath (remember, this will be very hot - use heatproof gloves). Using the mixer (on its slowest setting) or a whisk, blend the ingredients in the mixing container together until it has cooled to approximately 40-45°C (use the thermometer) - see fig 54: step 3.

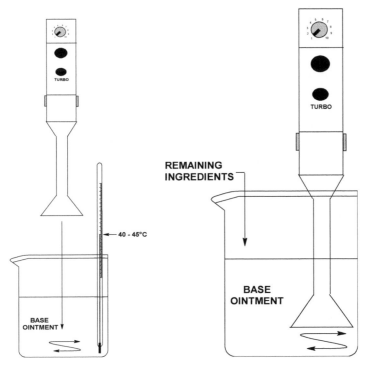

fig 54: step 3   fig 55: step 4

step 4

Once this temperature has been reached, you can add your remaining ingredients (i.e. TRs, MOs, CO_2 extracts), to achieve the end product desired (see fig 55: step 4).

Note: if you are adding EOs at this point, you may have to add a little extra to account for any loss due to evaporation at this temperature. This is unavoidable, as the ointment is solid below 25°C, making it impossible to blend in other ingredients.

Note: when adding your other materials, you need to reduce the amount of vegetable oil you use accordingly (e.g. if you add 30ml of an MO / IO, you need to reduce the amount of vegetable oil you use to 390ml to keep the ratios right).

Note: always remember to wipe any water droplets from the bottom of the mixing container after removing from the water bath, before the mix is poured into the storage containers. This is to ensure that no water is inadvertently added to the mix which could result in accelerated spoilage rates and a shorter shelf-life.

Note: it may be good idea to leave the water bath on a low heat after removing the mixing container for cooling. This is because some mixes can set quicker than others (depending on their ingredients and ratios) which can lead to difficulties when it comes to pouring the mix into the storage containers. If this happens you can then gently reheat the mix in the warm water bath (whilst stirring) until it is just liquid enough to make it easier to pour (try to avoid overheating).

Whilst the ointment mix is still warm and runny, you can pour it into your storage containers (do it quickly though, as it soon starts to set and become lumpy). Leave the containers to cool down before labelling and putting the lids on (formulation makes approx. 500ml of product).

Note: adding 0.5-1% vitamin E oil to the mix will help extend the ointment's shelf life.

balms

This is the same basic procedure and equipment as for making ointments. However, the following changes are made to the raw materials used: more solid fat (beeswax) and less liquid fat (vegetable oil, e.g. sweet almond oil) is used, and a vegetable butter (e.g. shea butter) is also added to the mix to give a smooth texture.

caution: this procedure requires the application of heat and the manipulation of hot wax / oil. You should therefore wear suitable eye and hand protection to avoid possible injury from burns.

Note: it will help you a lot if you can to get your sterile storage containers ready before you start this process.

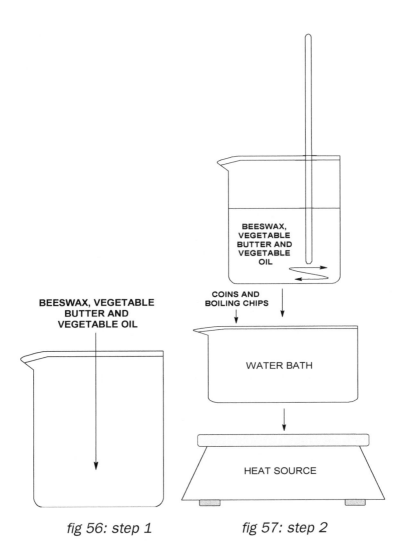

BEESWAX,
VEGETABLE
BUTTER AND
VEGETABLE
OIL

BEESWAX, VEGETABLE
BUTTER AND
VEGETABLE OIL

COINS AND
BOILING CHIPS

WATER BATH

HEAT SOURCE

fig 56: step 1   *fig 57: step 2*

step 1

Using the scales and measuring cylinder, weigh out 145g of beeswax and 110g of vegetable butter, and measure out 245ml of vegetable oil, and put these together into your mixing container (see fig 56: step 1).

step 2

On a heatproof work surface, set up your water bath on your hotplate; make sure that there is sufficient water present in your saucepan (not too much though!); and occasionally add more water when needed to avoid it boiling dry. Put the mixing container holding your ingredients from step 1 into the water bath. Turn on your hotplate, and, whilst stirring gently, heat the ingredients until the beeswax and vegetable butter melt into the vegetable oil (see fig 57: step 2).

Tip: use coins to lift the mixing container off the bottom of the saucepan, and use boiling chips.

step 3

Carefully remove the mixing container from your water bath (remember, this will be very hot - use heatproof gloves). Using the mixer (on its slowest setting) or a whisk, blend the ingredients in the mixing container together until it has cooled to approximately 40-45°C (see fig 58: step 3).

step 4

Once this temperature has been reached, you can add your remaining ingredients (i.e. TRs, MOs, CO_2 extracts), to achieve the end product desired (see fig 59: step 4).

Note: if you are adding EOs at this point, you may have to add a little extra to account for any loss due to evaporation at this temperature. As with ointments, this is unavoidable as the balm becomes solid below 25°C.

Note: when adding your other materials you need to reduce the amount of vegetable oil you use accordingly (e.g. if you add 30ml of an MO / IO, you need to reduce the amount of vegetable oil you use to 215ml to keep the ratios right).

Note: always remember to wipe any water droplets from the bottom of the mixing container after removing from the water bath, before the mix is poured into the storage containers. This is to ensure that no water is inadvertently added to the mix which could result in accelerated spoilage rates and a shorter shelf-life.

Note: it may be good idea to leave the water bath on a low heat after removing the mixing container for cooling. This is because some mixes can set quicker than others (depending on their ingredients and ratios) which can lead to difficulties when it comes to pouring the mix into the storage containers. If this happens you can then gently reheat the mix in the warm water bath (whilst stirring) until it is just liquid enough to make it easier to pour (try to avoid overheating).

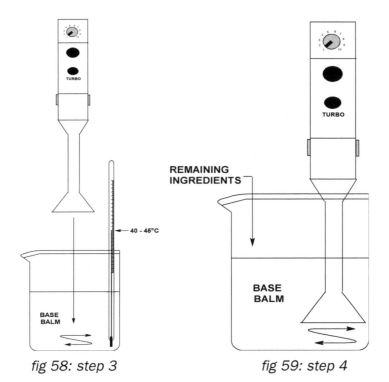

fig 58: step 3   fig 59: step 4

Whilst the balm mix is still warm and runny, you can pour it into your storage containers. Leave the containers to cool down before labelling and putting the lids on (formulation makes approx. 500ml of product). Again, try to do this quickly before the mix sets.

The resulting balm is waxier and less oily than the ointment (making it ideal for application to lips etc), and will also withstand higher temperatures before melting.

Note: adding 0.5-1% vitamin E oil to the mix will help to extend the balm's shelf life.

creams

To follow this procedure you will need the following equipment: scales; two measuring cylinders; two mixing containers; a water bath (i.e. a stainless steel saucepan with water in it); a two-ring electric hotplate; two polypropylene / glass stirrers; an electric hand-held mixer (or a whisk); a spatula; a thermometer (two would be better); ceramic tile (for cooling); and storage containers.

You will also require the following raw materials: glyceryl stearate; cetyl alcohol; a vegetable oil (or mix of oils, including MOs / IOs) of your choice; a vegetable butter of your choice (e.g. cocoa butter); sodium stearoyl lactylate; a water of your choice (in this example plain distilled water is used); glycerine; vitamin E oil (tocopherol); and any other raw materials you wish to add (e.g. EOs, aromatic extracts; preservatives etc).

caution: this procedure requires the application of heat and the manipulation of hot wax / oil. You should therefore wear suitable eye and hand protection to avoid possible injury from burns.

Note: it will help you a lot if you can to get your sterile storage containers ready before you start this process.

step 1

Using the scales, separately weigh out 15g of glyceryl stearate, 10g of cetyl alcohol, 5g of vegetable butter, and 22.5g of sodium stearoyl lactylate. Then use your measuring cylinders to separately measure out 70ml of vegetable oil (or mix), 377.5ml of plain distilled water, 15ml of glycerine, 10ml of tocopherol (vitamin E), and any other materials you wish to add to the mix.

Once this is done, all the ingredients for the fat phase (i.e. the glyceryl stearate, cetyl alcohol, vegetable butter, vegetable oil or mix), are put into a mixing container together. Then all the ingredients for the water phase (i.e. the sodium stearoyl lactylate, distilled water and glycerine), are put into a separate mixing container together (see fig 60: step 1).

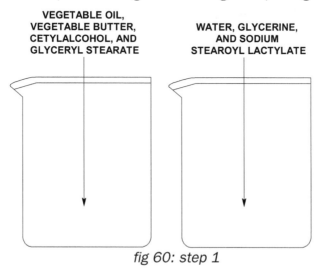

VEGETABLE OIL,
VEGETABLE BUTTER,
CETYLALCOHOL, AND
GLYCERYL STEARATE

WATER, GLYCERINE,
AND SODIUM
STEAROYL LACTYLATE

fig 60: step 1

Note: you should try to keep the mixer / whisk at the bottom of the mixing container (moving in a side-to-side motion) to avoid too much air being incorporated into the mix.

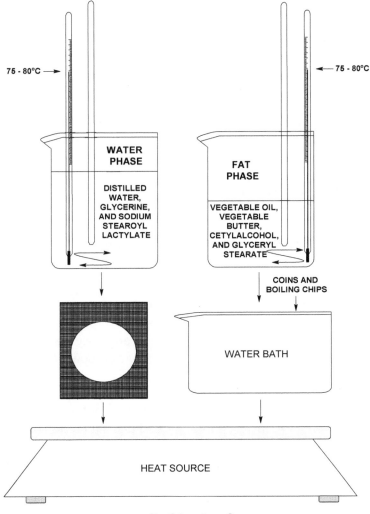

fig 61: step 2

step 2

On a heatproof work surface, set up your water bath on one of the rings of your hotplate (usually the bigger one if there are two sizes), making sure that there is sufficient water present in your saucepan (not too much!), and occasionally adding more water when needed to avoid it boiling dry).

Tip: as with ointments and balms, add coins & boiling chips.

Put the fat-phase mixing container from step 1 into the water bath. Then place the water-phase mixing container from step 1 on the other ring of the electric hotplate (remember to use the wire-screen ceramic heat dissipater). Turn on your hotplate (both rings), and, whilst stirring, gently heat the contents of both mixing containers to 75–80°C (make sure you don't overheat, as you may destroy the emulsifiers). See fig 61: step 2.

step 3

When both fat and water phases have reached 75-80°C, remove the thermometers, turn down the temperature of the fat-phase hotplate ring to its lowest setting, and turn off the water-phase hotplate ring. Now get your electric hand-held mixer (or whisk) ready, plugged in, and close to hand.

Then carefully (using heatproof gloves) pour the fat-phase mix into the water-phase mix via a thin, steady stream, whilst at the same time using the electric hand-held mixer on medium setting (or whisk) to continuously and vigorously blend the two phases together (see fig 62: step 3).

step 4

Whilst still blending the two phases together in the water-phase mixing container, transfer this mixing container (using heatproof gloves) into the water bath that the fat-phase container was originally heated in (which should still be warm, as the heat was only turned down and not off for this ring). See fig 63: step 4.

Continue to mix / whisk for a further 5 minutes, before turning off the hotplate, removing the mixing container and placing on a ceramic tile to allow your mix to cool to 40°C (use a thermometer to check this). You should still be mixing all this time (remember - try not to introduce any air!), but you can now set the mixer to a lower speed if you wish - depending on the desired thickness of your end product (the longer or faster it is mixed, the thicker the final product).

Note: you may then want to use a spatula to help scrape the sides if they become covered in mix.

Note: you can speed up the cooling process by immersing the mixing container in a large bowl of cold water (do not use ice as the mix may cool too rapidly and separate).

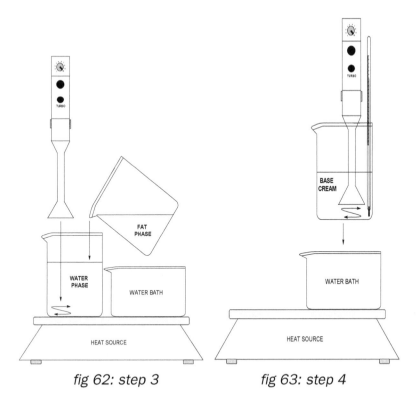

fig 62: step 3   fig 63: step 4

step 5

When the mix has reached 40°C you can add the ingredients that may otherwise be affected by the higher temperatures (e.g. preservative, vitamin E oil, natural moisturising factors, CO_2 extracts etc), whilst still mixing / blending your product (see fig 64: step 5).

Note: blend in these extra ingredients one at a time, adding the thicker materials first. You may use ice in your cooling water bath at this point.

step 6

When the mix has reached 25°C (use the thermometer to check this) you can add the remaining ingredients (e.g. EOs). See fig 65: step 6.

Whilst the mix is still warm you can pour your mix into your storage containers (formulation makes approx. 500ml of product).

Note: this recipe makes a regular cream suitable for normal skin types. For variations on this recipe for other skin types, please see appendix 9: cream recipe variation table.

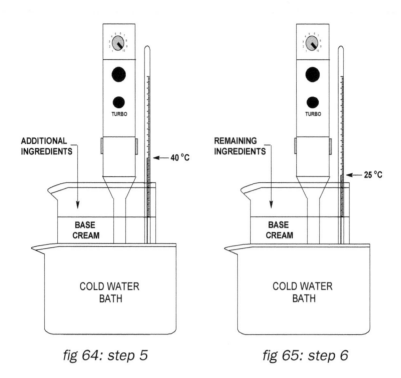

fig 64: step 5   fig 65: step 6

lotions

For lotions, we use the same basic procedure and equipment as for making creams. However, the following changes to the raw materials used are made: more water and less vegetable oil (e.g. jojoba oil), and a lower percentage of emulsifiers are used. Also, because of the higher water content in lotions, it is more likely that preservatives will need to be used, and in greater proportions.

caution: this procedure requires the application of heat and the manipulation of hot wax / oil. You should therefore wear suitable eye and hand protection to avoid possible injury from burns.

Note: it will help you a lot if you can to get your sterile storage containers ready before you start this process (as lotions are runnier than creams, they are usually stored in bottles rather than jars).

Note: you can also make lotions from creams (see page 139: 'how to make lotions easily from creams').

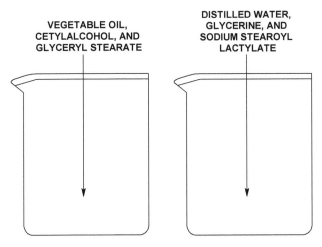

fig 66: step 1

step 1

Using the scales, separately weigh out 5g of glyceryl stearate, 5g of cetyl alcohol, and 12.5g of sodium stearoyl lactylate. Then use your measuring cylinders to separately measure out 40ml of vegetable oil(s), 437.5ml of plain distilled water, 10ml of glycerine; 10ml of vitamin E oil, and any other materials you wish to add to the mix.

Once this is done, all the ingredients for the fat phase (i.e. the glyceryl stearate, cetyl alcohol, and vegetable oil) are put into a mixing container together. Then all the ingredients for the water phase (i.e. the sodium stearoyl lactylate, distilled water and glycerine) are put into a separate mixing container together (see fig 66: step 1).

step 2

On a heatproof work surface, set up your water bath on one of the rings of your hotplate (usually the bigger one if there are 2 sizes), making sure that there is sufficient water present in your saucepan (not too much!), and occasionally adding more water when needed to avoid it boiling dry). Put the fat-phase mixing container from step 1 into the water bath.

Tip: as with creams, add coins and boiling chips to avoid bumping during heating.

Then place the water-phase mixing container from step 1 on the other ring of the electric hotplate (remember to use the wire-screen ceramic heat dissipater).

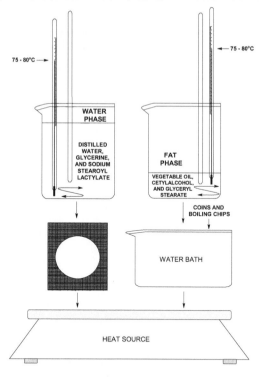

fig 67: step 2

Turn on your hotplate (both rings), and, whilst stirring, gently heat the contents of both mixing containers to 75–80°C (use the thermometers to check this). As with creams, make sure not to overheat. See fig 67: step 2.

step 3

When both fat and water phases have reached 75-80°C, remove the thermometers, turn down the fat-phase hotplate ring to its lowest setting, and turn off the water-phase hotplate ring. Now get your electric hand-held mixer (or whisk) ready, plugged in, and close to hand.

Then carefully (using heatproof gloves) pour the fat-phase mix into the water-phase mix via a thin, steady stream, whilst at the same time using the electric hand-held mixer on its lowest setting (or whisk) to continuously and vigorously blend the two phases together (see fig 68: step 3).

Note: you should try to keep to mixer / whisk at the bottom of the mixing container (moving in a side-to-side motion) to avoid too much air being incorporated into the mix.

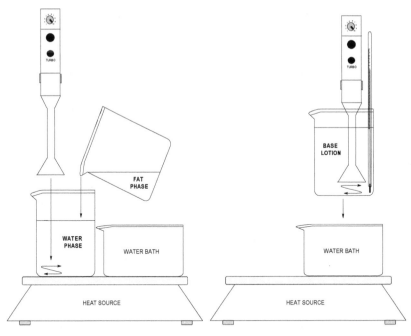

fig 68: step 3   fig 69: step 4

step 4

Whilst still blending the two phases together in the water-phase mixing container, transfer this mixing container (using heat proof gloves) into the water bath that the fat-phase container was originally heated in (which should still be warm, as the heat was only turned down and not off for this ring). See fig 69: step 4.

Continue to mix / whisk your mix for a further 5-10 minutes, before turning off the hotplate, removing the mixing container and placing on a ceramic tile to allow your mix to cool to 40°C (use a thermometer to check this). You should still be periodically mixing (remember - try not to introduce any air), but you can set the mixer to a lower speed if you wish - depending on the desired thickness of your end product (remember, if you mix too long or too fast, the lotion may become too thick).

Note: you may then want to use a spatula to help scrape the sides if they become covered in mix.

Note: you can speed up the cooling process by immersing the mixing container in a large bowl of cold water (do not use ice as the mix may cool too rapidly and separate).

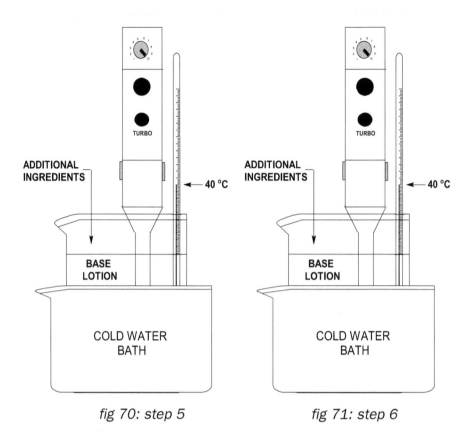

fig 70: step 5   fig 71: step 6

step 5

When the mix has reached 40°C you can add the ingredients that may otherwise be affected by the higher temperatures (e.g. preservative, vitamin E oil, natural moisturising factors, CO_2 extracts etc), whilst still mixing / blending your product (see fig 70: step 5).
Note: blend in these extra ingredients one at a time, adding the thicker materials first. You may use ice in your cooling bath at this point.

step 6

When the mix has reached 25°C (use the thermometer to check this) you can add the remaining ingredients (e.g. EOs). See fig 71: step 6. Whilst the mix is still warm you can pour you mix into your storage containers (formulation makes approx. 500ml of product).

trouble-shooting

The basic formulas presented in this book will allow you to get to grips with the processes, equipment, and raw materials involved in base product manufacture. However, it is the ultimate aim of this book to enthuse you to adapt these formulas to meet your own requirements.

I have tried to outline all the steps involved in each procedure in detail, in order to make them as easy as possible to follow. Even so, from time to time, you may run into some problems when following these formulas (especially when doing them for the first time). If this does happen I suggest that you try to retrace your steps to make sure that the right materials were added to the right phases at the correct temperatures (e.g. it is common for people to add the wrong emulsifiers to the wrong phases, or for the water phase to be added to the oil phase rather than the other way round). These mistakes may not just happen to beginners, but also to people experienced in making their own products who may be in a rush, or distracted etc.

It is for this reason that I have added a section here, in order for people to be able to diagnose and remedy some of the more common problems that may occur when making your own base products.

customising your formulas

The first thing you need to do is decide what type of end product you want (i.e. what is the product going to be used for), and what raw materials you want to include (e.g. materials that help moisturise the skin) as well as what raw materials you do not want to include (e.g. any preservatives or materials that are contraindicated or that you may be allergic to).

When you have worked out what you want in your product then you need to produce a formula that you can work by, that details the ratios of the materials you are going to use in respect to each other. The easiest way to do this is to work to a formulation that gives you 100ml or 100g, which will then mean that every 1ml or 1g will be equal to 1% of your total mix. From here you can scale the formula up to however much you need to make.

When you add more products to the recipe, you need to adjust the amounts of the other ingredients accordingly (e.g. the more water-based products that are added, the less distilled water is used; and the more oil-based products are added, the less oils / fats are used - and the more emulsifiers are added to accommodate this). In general, for creams, the amount of sodium stearoyl lactylate is roughly equal to the amount of glyceryl stearate and cetyl alcohol added together, to keep the ratios right. Remember: the maximum amount of oils / fats used in creams or lotions should not exceed 40% of the mix otherwise the cream or lotion will separate out.

To make thicker gels, just increase the percentage of xanthan gum used; to make harder ointments or balms, increase the percentage of solid fats (e.g. beeswax) used.

Note: it is easier to start with adjustment of the fat phase first and then work out how much you need in the water phase accordingly.
Also, see appendix 9: cream recipe variation table.

products that are lumpy or grainy

This can occur when making any of the base products covered and is mostly due to one of two reasons. Firstly, some of the materials present may not have melted sufficiently (e.g. the solid fats in the fat phase) - in which case you can reheat the mix to 75°C (as long as the heat-sensitive materials have not already been added). Secondly, some of the materials may not have been dissolved / mixed in efficiently (e.g. the emulsifiers or thickeners) - in which case if you used a whisk to mix your ingredients, then you now need to use an electric hand-held mixer (although do not mix for too long or the mix may become too thick in the case of creams or lotions, or may turn white and stiff, in the case of gels). Make sure that any ingredients that may have become stuck to the inside of the mixing container are thoroughly mixed in.

This can also happen with creams or lotions if the fat phase and water phase temperatures are too different at the point of mixing (both should be 75-80°C at the point of blending).

products that are too heavy

This can happen with creams or lotions, and can be remedied by adding a little more water to the product (as long as it doesn't make it too runny). Remember to increase the percentage of preservative accordingly if you do this. Alternatively, you can reduce the amount of water that is used in the water stage by 10-25% (at the start), and then mix in 10-25% aloe vera gel (see appendix 6: recommended raw materials), when the temperature of the mix has reached 35-40°C, resulting in a lighter, 'fluffier' cream.

Note: the aloe vera gel that you use needs to be of a thin / pourable consistency, otherwise when you add this to the mix and blend it in you may end up with a product that has the consistency of chewing gum.

products that are too thin or too thick

These problems mostly happen in the production of creams or lotions. If the product is too runny or thin it could be due to any of the following: too little sodium stearoyl lactylate was added at the water phase (or it was forgotten altogether, or not mixed in thoroughly); the fat and water phases were not both at 75-80°C when mixed together; the fat phase was added too quickly to the water phase; the water phase was added to the fat phase by mistake; the wrong materials were added to the wrong phase; or the product was cooled down too quickly. To remedy this (as long as the above emulsifier has been added in the right proportion to the right phase), you can try re-heating the product to 75-80°C in a water bath and use an electric hand-held mixer (on a faster speed) for approximately 5 minutes to obtain a thicker, creamier product - and then allow the product to cool slowly.

Note: you can also sprinkle a little xanthan gum (0.5-1%) into the mix as you blend it, to thicken it up a bit more if needed.

If the product is too thick or hard, this can be due to too much solid fat being incorporated into the mix (adjust the mix accordingly next time), or if the water and fat phases were mixed together too vigorously or for too long. To remedy this, cold water (preferably pre-sterilised / distilled) can be stirred into the mix (a little at a time) whilst using an electric hand-held mixer to continuously mix (at low speed) the ingredients until the right consistency is reached.

products that separate

If a lotion starts to separate, this can be because materials have been left out (e.g. one or both of the emulsifiers), or because of inadequate mixing of the ingredients (e.g. if they were added to the wrong phases, if the fat phase was added too quickly to the water phase, if the phases were not added at the right temperature, if the water phase was added to the fat phase, or if the product was allowed to cool too quickly or not stirred at room temperature). To remedy this (as long as both emulsifiers have been added in the right proportion to the right phases), you can try re-heating the product to 75-80°C in a water bath, and then use an electric hand-held mixer to blend for approximately 5 minutes, before allowing to cool again.

products with bad odour

This may be because the fats (either liquid or solid) that you have used may be rancid (this is common when using wheatgerm oil) or that they may naturally have a strong odour (e.g. borage MO / IO has a slightly mouldy smell to it). If the smell is due to rancid oils it is best to throw the product away. Next time you make the product, choose a different fat that has similar properties but is less susceptible to oxidation and may therefore impart a more discreet odour - for example, add 0.5-1% pure vitamin E oil (tocopherol) instead of wheatgerm oil.

Alternatively, you could try to mask the smell by using DWs / EOs, and / or CO_2 extracts (as long as the smell is not due to rancid oils).

problems with the colour of products

This can sometimes happen if decocted / infused / macerated / percolated waters are incorporated into the mix, which can often impart an unpleasant greyish-green colour to your final product. To remedy this, you can add a few drops of carotene to the mix during the fat phase, which will give your product a yellow colour (depending on how strong the original colour was). Alternatively, you can add a little German chamomile EO to give a blue tint, St John's wort MO to give a red tint, marigold MO to give a golden tint, or rose petal MO to give a pink tint. CO_2 extracts may also be used to help bring a particular colour to a product (although you may want to do a test before adding to the whole batch).

Note: if the product is exposed to direct sunlight for too long then the colour from the carotene can fade.

how to make lotions easily from creams

Once you have made a cream it is very easy to then convert it into a lotion.

In order to do this you need to slowly add the same volume of water (or equivalent, preferably pre-sterilised / distilled) to the amount of cream you have (e.g. if you have 500ml of cream then add another 500ml of water). As you do this you should use an electric hand-held mixer (set on a low speed), to mix in the water.

Note: any EOs you want to be incorporated into the lotion should already be added to the cream before conversion to a lotion, otherwise they may separate out when the cream becomes a lotion.

As this product will now contain a higher percentage of water, its shelf life will be considerably shortened unless a higher percentage of preservative is added (usually doubled if water content is doubled). Alternatively, the lotion can be made with minimal use of preservatives if carefully packed into a sterilised airless pump container in a clean fridge, and as long as it is used within 2 days and does not contain decocted / infused / macerated / percolated waters).

This also applies if water-based herbal extracts are used (such as decocted / infused / macerated / percolated waters), where an even greater percentage of preservative may be required, otherwise the shelf life will be considerably reduced.

You may also want to add some more moisturising ingredients so that the lotion is not too watery (e.g. you could add an extra 2% glycerine and 2% natural moisturising factors, as the water is being blended in).

When the right consistency has been reached, the product can be bottled and labelled, before storage.

Note: if euxyl K700 is used, then you have to test / adjust the pH of the final product to 5.5 or below to activate it.

storage & shelf-life of skin-care products

Most of this section has already been explained and covered in part 1 of this book (see page 71: 'storage and shelf-life'). However, there are a few further points that should be mentioned here.

degradation

As with EOs, some of the skin-care products discussed can also be affected by the presence of moisture.

For example, to stop biological contamination (see 'contamination' section below), often a product contains preservative (or a mix of preservatives), added in proportion to the amount of water present in a product (e.g. you would use more preservative in a lotion than you would in a cream). If the product is then stored in conditions that allow more moisture to enter, this may increase the likelihood of biological contamination, as there would be insufficient preservative present to counteract this. For example, if a product is kept in a bathroom (especially if the lid was left off for a while); or if the jar was left in direct sunlight, and condensation formed on the lid - then moisture could sit on the surface of the product and allow mould to grow.

contamination

The biggest problem you will find when making your own base products is contamination (especially from organisms). A lot of people (quite reasonably) want to make their own products without using the harsh preservatives that are commonly used in commercial products. However, if you choose to go down this route, you need to be aware that there are important factors to take into consideration - some of which will be discussed below and some of which have already been discussed in the 'preservatives' section on page 100.

When making your own base products, contamination can fall into three main areas: contamination of starting material, contamination from organisms, and contamination from other products.

contamination of starting materials

It is important to know where the raw materials for your products have come from and (if possible) how they were made. (Traceability is becoming more important in terms of natural skin-care products.) This is very important because some raw materials may have been obtained via ethically or ecologically unsound processes (in most cases botanical, semi-synthetic, or synthetic alternatives can be used). Ultimately, your final product is only as good as the starting materials it was made from.

contamination from organisms

This occurs more with gels, creams or lotions than ointments or balms (this is because the former products contain higher ratios of water than the latter).

The most common types of organisms that can be present are bacteria and fungi (especially moulds), and problems can occur if utensils or storage containers are not properly cleaned / sterilised before use or reuse.

The following advice is therefore given to help prevent this type of contamination:

- bottle up your products as soon as possible after they have been made.
- try to keep all equipment and work spaces as clean as possible, to try and prevent bacteria / fungi being introduced into your mixes (especially if you choose not to use preservatives); this can be achieved by habitually cleaning all equipment meticulously both before and after use, and sterilising utensils (by boiling in water for at least 10 minutes).
- always follow GMP (good manufacture practice – see *resources*) standards by always wearing 'indoor' shoes, an apron, and head gear (so that no hairs fall into your product) when you are making your products.
- always use an alcohol-based sterilising spray (must be at least 25% alcohol) on all utensils (use a paper towel to wipe them dry) and on your hands before making your products (remembering to re-apply after you go to the toilet or if you wash your hands).
- it is also advantageous to divide your work area up into smaller separate 'stations' (e.g. an area for storage, an area for making the products, an area for bottling, and an area for packing), that flow around the work area in one direction (e.g. clockwise around the

room) from one station to another. In this way you won't be randomly moving around your work area, and so you will minimise the possibility of any cross-contamination from one batch / product to another (especially if there are several people in the work area).

- you may want to purchase a HEPA (high-efficiency particulate absorbing) air filter (used in labs and hospitals) to help reduce the levels of airborne biological contaminants in your work area.
- try to keep your operation as small as possible for your needs, only making as much product as you need for the next few weeks, and always keeping your products in a thoroughly clean refrigerator (preferably one that is used just for products, and not for food as well). In this way you can minimise the use of preservatives, as long as the product is used up very quickly. Treat it as you would fresh dairy produce, and you will always be using good fresh products.
- you may also consider the type of storage container you will use - for example tubes of product may last longer than jars because a smaller surface area is exposed to atmospheric contamination from organisms every time the product is dispensed. If you are using jars on the other hand, then it is advantageous to use small disposable spatulas to dispense the product rather than using your fingers, in order to reduce the likelihood of contamination.
- do not use (in fact, throw out) any product that looks like it has been contaminated.
- use products for external application only, and do not return any unused product back into its storage container.

contamination from other products

This has already been covered in part 1, on page 72.

labelling

One last thing to consider is labelling your product. This may seem obvious but it can easily be missed if you make a number of products at once, and have to guess which product is in which container. It is also important from a safety point of view (e.g. you should always write 'keep out of reach of children' and 'for external use only' on your product containers).

I find the following useful to include: raw materials used (in descending percentage order); the date the product was made (along with the date the product should be used by); a batch number; how it should be stored; and who it was made by (if in a team).

Note: if you want to sell your products to the general public then you should be aware that there are stringent guidelines on how the product should be packaged and labelled, and that it should also be independently tested for safety before it is sold openly (see *resources*). An-up to-date version of these guidelines can be freely obtained from your local Trading Standards office.

appendix 1:
the distillation kit

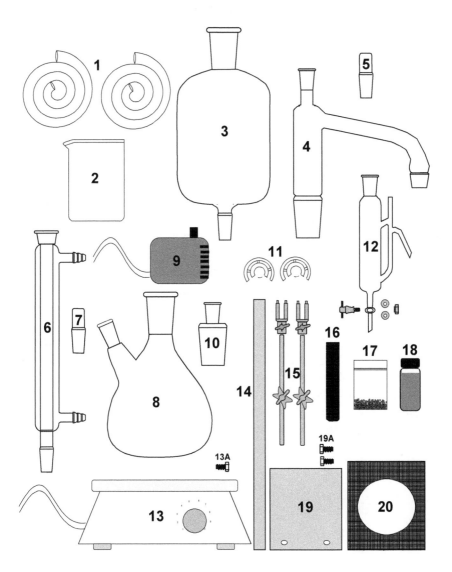

fig 72: the distillation kit - see next page for key

1. 2 x five-foot hoses – for condenser water
2. Collection beaker (Pyrex®; 1-litre with graduations) – to collect distillate water produced
3. Biomass flask (Pyrex®; 2-litre) - to house herb material for distillation
4. Still head (Pyrex®; with B40/38 cone, S29 ball joint, and B24/29 socket for stopper or thermometer holder)
5. Stopper for socket of still head (Pyrex®; with B24/29 cone)
6. 'West' condenser (Pyrex®; with S29 cup joint and B24/29 cone)
7. Stopper for side arm of boiling flask (Pyrex®; with B24/29 cone)
8. Boiling flask (Pyrex®; 2-litre; with graduations, B40/38 socket and side arm with B24/29 socket)
9. Water pump – to circulate iced water through condenser
10. Socket reduction adaptor for boiling flask (Pyrex®; B24/29 socket to B40/38 cone)
11. 2 x 'Keck' clips – (red and green) to secure ground glass joints together
12. Specially-designed receiver / separatory funnel (Pyrex®; 30ml; with B24/29 socket and Teflon® stopcock)
13. Hotplate - 1.5 kilowatt, with built-in apparatus support rod holder and securing screw (13A)
14. Apparatus support rod - to secure apparatus in place
15. 2 x laboratory clamps – to secure apparatus
16. Stainless steel mesh screen filter – to secure herb material in biomass flask
17. Bag of boiling chips (10g) - to stop 'bumping' when boiling water
18. Tub of silicone grease (10g) - to lubricate glass ground joints
19. Metal heat shield with two securing screws (19A) – to protect essential oil collected from excess exposure to heat
20. Wire-screen ceramic heat dissipater – to allow even heat transference to boiling flask

appendix 2:
distillate water monographs

Note: many books contain essential oil monographs (see *resources*), but distillate water, macerated / infused oil and herbal tincture monographs are harder to find. The appendices that follow contain monographs for a selection of commonly used medicinal plants.

Achillea millefolium: yarrow

family
Asteraceae (formerly *Compositae*)

part used
Flowering tops (fresh or dried).

type of distillate water
Can be either AW or hydrosol, depending on the type of distillation used.

constituents
The EO contains: camphor; sabinene; 1,8-cineole; -pinene; ß-pinene; camphene; and azulenes (converted to chamazulene during steam distillation), some of which may also be dispersed within the DW.

properties and uses (external)
Anti-inflammatory; anti-allergenic; antiseptic; astringent; and haemostatic / styptic.
Add to a 'Sitz bath' for haemorrhoids or postpartum healing.
Combined with other anti-inflammatory herbs / products (e.g. German chamomile DW) for inflamed / irritated skin conditions as part of a lotion or compress.
Combine with peppermint water as a styptic aftershave toner.

dosage
Add up to 20% in creams / lotions in combination with other astringent herbs / products - e.g. witch hazel (*Hamamelis virginiana*) DW for weeping conditions, wounds, and thread / varicose veins.

contraindications
Do not use in cases of pregnancy (and breast feeding), children under 2 years of age, epilepsy, and kidney disease due to possible presence of thujone.
May cause sensitisation / phototoxicity in some individuals.

Chamomilla / Matricaria recutita: German chamomile

family
Asteraceae (formerly *Compositae*)

part used
Flowers (fresh or dried).

type of distillate water
Can be either AW or hydrosol, depending on the type of distillation used.

constituents
The EO contains: proazulenes; bisabolol; farnescene; pinene; anthemal; spiroether; angelic acid; tiglic acid; malic acid; bitter glycosides; and coumarins, some of which may also be dispersed within the DW.

properties and uses (external)
Anti-inflammatory; anti-allergenic; vulnerary; and antiseptic.
It is indicated for any inflamed / irritated skin conditions (such as nappy rash; eczema; urticaria; acne rosacea/vulgaris; and varicose ulcers). Combine with chickweed (*Stelleria media*) cream, and lavender (*Lavandula officinalis*), or damask rose (*Rosa damascena*) DWs for itchy / inflamed conditions.

dosage
Add up to 20% in creams / lotions, or spray liberally on any inflamed / irritated skin condition.
Appliy to a warm bath just before entering (approx 50ml) for its soothing / relaxing effects.

contraindications
There are no known contraindications for this DW.

Citrus aurantium var. amara: bitter orange (flower)

family
Rutaceae.

part used
Flowers (fresh).

type of distillate water
AW.

constituents
The EO contains: - and ß-pinene; camphene; -terpinene; nerol; neryl acetate; farnesol; geraniol; linalool; nerolidol; methyl anthranilate; and indole, some of which may also be dispersed within the DW.

properties and uses (external)

Sedative; tranquilliser; antidepressant; astringent; and antiseptic.

It is thought to improve the skin's micro-circulation and promote cellular regeneration.

It is suitable for skin that is inflamed, sensitive, or prone to broken capillaries.

Combine with rose geranium (*Pelargonium graveolens*) DW as a facial toner for oily / sensitive skin.

dosage

Like German chamomile (*Chamomilla recutita*) DW, this DW can be applied to a warm bath just before entering (approx 50ml), to relax the body, calm the mind and lift the spirits.

Add up to 20% in creams / lotions (good for oily skin).

contraindications

Generally non-toxic / non-irritating. However, it can sometimes be too drying for people with already dry skin.

Foeniculum vulgare: fennel

family

Apiaceae (formerly *Umbelliferae*).

part used

Seed (dried).

type of distillate water

Can be either AW or hydrosol, depending on the type of distillation used.

constituents

The EO contains: trans-anethol; fenchone; methylchavicol; and - pinene, some of which may also be dispersed within the DW.

properties and uses (external)

Hydrating and softening to the skin.

Use in a spritzer as a hydrating facial toner.

Use in creams for dry, ageing skin and vaginal dryness.

dosage

Add up to 10% to creams / lotions.

contraindications

Do not exceed recommended dose.

Do not use in cases of pregnancy, epilepsy, oestrogen-dependent cancers and endometriosis.

Some individuals may experience sensitisation to this DW when used externally.

Do not use on sensitive or damaged skin.

Hamamelis virginiana: witch hazel

family
Hamemilacea.

part used
Leaves, twigs and bark (fresh or dried).

type of distillate water
HW

constituents
The constituents of the DW are as of yet not well known.

properties and uses (external)
Astringent; anti-oxidant; anti-inflammatory; cicatrisant; and antiseptic.
Excellent for: burns (including sunburn); swellings; weeping eczema;
varicose / thread veins; phlebitis; varicose ulcers; bed sores; bruises;
sprains / muscle strains; and insect bites. Combine with German
chamomile (*Chamomilla recutita*) DW to soothe inflamed skin.
Use as part of a compress (esp. for sprains / muscle pains).
Use in creams or ointments for haemorrhoids.
Use in lotions as a toner for oily / lax skin, or dabbed around the eyes
to refresh them and reduce puffiness.
Combine with rose geranium (*Pelargonium graveolens*) or bitter orange
flower (*Citrus auranthium var. amara*) DWs as a daily astringent facial
toner for oily / blemished skin.

dosage
Add up to 20% to creams / lotions (especially for damaged / ageing skin).

contraindications
There are no known contraindications if used externally.

Lavandula officinalis: true lavender

family
Lamiaceae (formerly *Labiateae*).

part used
Flowering tops (fresh or dried).

type of distillate water
Can be either AW or hydrolat, depending on the type of distillation used.

constituents
The EO contains: linalyl acetate; lavanduyl acetate; linalool; terpineol;
cineol; borneol; nerol; camphor; limonene; cardinene; cardophyllene,
some of which may also be dispersed within the DW.

properties and uses (external)
Antidepressant; relaxant; analgesic; vulnerary; antiseptic; cicasitrant; and insecticidal.
Incorporate into a compress for headaches / migraines, rheumatic pain, and neuralgia.
Use for all skin types as a spray for hot flushes, as a daily skin toner, for burns (including sun / razor burn (combines well with witch hazel - *Hamamelis virginiana* DW for this), wounds, nappy rash, insect bites (combines well with German chamomile - *Chamomilla recutita* DW), and almost any skin problem.

dosage
Like German chamomile *(Chamomilla recutita)* DW, this DW can be applied to a warm bath just before entering (approx 50ml), to relax the body and calm the mind (especially good for restless / irritable children). Add up to 20% to creams / lotions (good for eczema).

contraindications
Generally non-toxic / non-irritating, although there have been rare cases of people developing skin sensitisation to this DW.

Mentha x piperita: peppermint

family
Lamiaceae (formerly *Labiateae*).

part used
Leaves and tops (fresh, wilted or dried).

type of distillate water
Can be either AW or hydrolat, depending on the type of distillation used.

constituents
The EO contains: menthol; menthone; piperitone; cineole; menthyl; 1,8-cineole; methyl acetate; methofuran; isomethone; limonene; and - and ß-pinene, some of which may also be dispersed within the DW.

properties and uses (external)
Refrigerant; anaesthetic; anti-inflammatory; antiseptic; and anti-parasitic.
Use for hot, itchy rashes (e.g. razor burn) or insect bites.
Use as a compress for muscle and nerve pain, bruises and contusions, or added to a cream as an 'after exercise' rub for sore / tired muscles.

dosage
Add up to 10% in creams / lotions.
Apply liberally via spray.

contraindications

Generally non-toxic / non-irritating. However, use of this DW should be avoided in large amounts during pregnancy or whilst breast-feeding, and it should not be used on children under 5 years old.

Some skin types may be sensitive to this DW.

Pelargonium graveolens: rose geranium

family
Geraniacea.

part used
Leaves and flowering tops (fresh or dried).

type of distillate water
Can be either AW or hydrosol, depending on the type of distillation used.

constituents
The EO contains: citronellol; geraniol; linalool; citronellyl formate; and geranial, some of which may also be dispersed within the DW.

properties and uses (external)
Anti-depressant; astringent; styptic; cicatrisant; vulnerary; antiseptic; insect repellent; anti-inflammatory; and analgesic.

Suitable for all skin types (including sensitive skins); balances sebum production (good for acne and oily skins), good for toxic skin conditions; wounds; blemishes; scars; or as a daily skin toner.

Use as a cream for haemorrhoids and varicose / broken veins, or as a compress or cream for nerve pain or hot / inflamed joints and muscles. Use as a spritzer for hot flushes (combines well with damask rose - *Rosa damascena* DW for this), or use neat as a makeup remover.

dosage
Like German chamomile (*Chamomilla recutita*) DW, this DW can be applied to a warm bath just before entering (approx 50ml), to relax the body, calm the mind and lift the spirits.

Add up to 20% to creams/lotions.

contraindications
There are no known contraindications for this DW.

Rosa damascena: damask rose

family
Roseacea.

part used
Petals (fresh or dried).

type of distillate water
AW.

constituents
The EO contains: geraniol; nerol; citronellol; stearpoten; phenyl ethanol; farnesol; eugenol; geranic acid; and myrcene, some of which may also be dispersed within the DW.

properties and uses (external)
Calming and uplifting; anti-inflammatory; antiviral; cooling astringent; and helps to restore the skin's natural pH.
Suitable for all skin types (especially dry, mature, sensitive skins).
Use in creams for eczema.
Use as spritzer for menopausal hot flushes, or as a toner for dry / mature skin.

dosage
Add up to 20% to creams / lotions.

contraindications
There are no known contraindications for this DW.

Salvia triloba: Greek sage

family
Lamiaceae (formerly Labiateae).

part used
Leaves and tops (fresh or dried).

type of distillate water
Can be either AW or hydrosol, depending on the type of distillation used.

constituents
Similar to Spanish sage (Salvia lavendlaefolia). The EO contains: cineole; linalool; -pinene; camphene; camphor; borneol; and thujone (0.72-1.86%), some of which may also be dispersed within the DW.

properties and uses (external)
Outstanding anti-microbial action - especially against klebsiella, streptococcus / staphylococcus, and Candida albicans (thrush) infections, and anti-inflammatory.

Use as a cream / lotion against fungal infections (e.g. thrush); for boils / ulcers / wounds; for muscle aches / pains; and for acne.

Use alone as a hair tonic / shiner. Rub into the scalp, and then spray over wet hair after the final rinse (can also help with dandruff).

dosage

Add up to 15% to creams / lotions.

contraindications

Do not use during pregnancy, or with epileptics

appendix 3: macerated / infused oil monographs

Arnica montana: arnica

family
Asteraceae (formerly *Compositae*).

part used
The whole plant is used, either fresh or dried to make either an MO or IO.

constituents
EOs (including triterpenoid alcohols - arnidiol and faradiol, thymol esters, and sesquiterpene arnicolide); flavonoid heterosides (e.g. kempferol and quercetol); malic acid; tannic acid; arnicin; sesquiterpene lactones; astragalin; isoquercitrin; and various trace compounds.

properties and uses (external)
Analgesic; anti-inflammatory; styptic; and vulnerary.
This plant has been used for centuries for muscle and joint pain / inflammation, and is especially indicated for bruises and sprains.
You also have the properties of the chosen base oil that was used to make the MO / IO.

dosage
Incorporate into massage blends (3-10%) along with base oil(s) and EOs. Incorporate into gels, ointments / balms, creams / lotions in place of the vegetable oil(s) used in the fat phase (up to 30%).

contraindications
Some people can be sensitive to this plant (may cause skin inflammation), especially if used over long periods of time.
caution: this MO / IO should only be used externally, as internal application could result in cardiac arrest!

Calendula officinalis: marigold

family
Asteraceae (formerly *Compositae*).

part used
The flowers (either fresh or dried) are usually macerated in sunflower oil (organic) in strong sunlight for several weeks to produce a golden-coloured MO.

constituents

Triterpenoid saponins; carotenoids (e.g. carotene, calendulin, and lycopin); bitters; sterols; flavonoids; resins; mucilage; EOs; palmic acid; malic acid; salicylic acid; and various trace minerals.

Note: if fresh blossoms are used then small amounts of phyto-oestrogens may also be present in the MO.

properties and uses (external)

Antiseptic; anti-inflammatory; astringent; cicatrisant; styptic; and vulnerary.

An excellent general tonic for the skin, which is also mild enough to use in baby skin care preparations. Also good for dry / cracked skin, especially lips (good for lip balms).

You also have the properties of the chosen vegetable oil that was used to make the MO.

dosage

Incorporate into massage blends (3-10%) along with base oil(s) and EOs. Incorporate into gels, ointments / balms, creams / lotions in place of the vegetable oil(s) used in the fat phase (up to 100%).

contraindications

There are no known contraindications for this MO / IO.

Harpagophytum procumbens: devil's claw

family

Pedaliaceae.

part used

The dried herb is used to make either an MO or IO.

constituents

Phytosterols; flavonoids (including kaempferol, kaempferilde, fisetine, and luteoline); fatty acids; triterpenic acids / esters (e.g. ursolic and oleanolic acids / esters); phenolic acids (e.g. cinnamic and cholorogenic acids); sugars (e.g. glucose, fructose, raffinose, and stachyose).

properties and uses (external)

Analgesic; anti-inflammatory; and antispasmodic.

This MO / IO helps to relieve the pain associated with a wide range of muscular and joint problems (especially good for rheumatic and arthritic conditions).

You also have the properties of the chosen vegetable oil that was used to make the MO / IO.

dosage

Incorporate into massage blends (3-10%) along with base oil(s) and EOs. Incorporate into gels, ointments / balms, creams / lotions in place of the vegetable oil(s) used in the fat phase (up to 50%).

contraindications

This MO / IO should be used externally only.

Hydrocotyl asiatica: gotu kola

family

Apiacea (formerly *Umbelliferae*).

part used

The dried, whole herb is used to make either an MO or IO (usually using sweet almond oil as the base).

constituents

Glycosides (e.g. centelloside, and asiaticoside); alkaloid (hydrocotylin); bitter (vellarine); triterpenic acids (including asiatic acid), madecassic acid, hydrocyanic acid); resins; tannins; and trace amounts of catechol and epicatechol.

properties and uses (external)

Antiseptic; anti-inflammatory; antipyretic; astringent; venous decongestant; and vulnerary.

This MO / IO has excellent applications for wound healing including dermatitis, pruritus, scarring, boils, acne, varicose veins, leg ulcers and haemorrhoids.

You also have the properties of the chosen vegetable oil that was used to make the MO / IO.

dosage

Incorporate into massage blends (3-10%) along with base oil(s) and EOs. Incorporate into gels, ointments / balms, creams / lotions in place of the vegetable oil(s) used in the fat phase (up to 70%).

contraindications

This MO / IO is generally regarded as being non-toxic / non-irritating, although when applied internally as a herbal medicine, caution is advised about prescribing this herb to people with epilepsy (due to its alkaloid content), and so this is not recommended until more is known.

Hypericum perforatum: St. John's wort

family
Hypericaiceae.

part used
The flowers (partially dried) are usually macerated in either virgin olive oil (preferably extra virgin and organic) or sunflower oil (organic) in strong sunlight for several weeks to produce a blood-red coloured MO.

constituents
Hypericin; flavonoids (e.g. hyperocide, quercetin, quercitrin, and rutin); catechic tannins; EOs (including pinene, limonene, caryophyllene, humulene, and hypericine).

properties and uses (external)
Antispasmodic; astringent; analgesic; antiviral; and vulnerary.

This MO is exceptionally beneficial when applied to wounds where there is either inflamed nerves or nerve tissue damage present (e.g. sciatica, neuralgia, and fibrositis), and joint pain (e.g. rheumatism and gout).

It can also be useful in skin conditions (e.g. ulcers, wounds, urticaria, vitiligo, and herpes), and can help to tighten / tone the skin too. It also makes an excellent shaving oil (massage into the area before applying the shaving foam), due to its soothing, and analgesic effects.

You also have the properties of the chosen vegetable oil that was used to make the MO.

dosage
Incorporate into massage blends (3-10%) along with base oil(s) and EOs. Incorporate into gels, ointments / balms, creams / lotions in place of the vegetable oil(s) used in the fat phase (up to 50%).

contraindications
This MO / IO is generally regarded as being non-toxic / non-irritating, although the pure EO of this plant has been known to cause phototoxicity, so due care should be taken when using the MO (i.e. avoid exposure to direct sunlight / sun beds for 12 hours following application).

appendix 4: herbal tincture monographs

Aesculus hippocastanum: horse chestnut

family
Hippocastanacea.

part used
Mostly the fruit (nut) is used for external application.

constituents
Contains hippocaesculin; triterpenoid saponins (e.g. aescin); coumarins; and flavonoids.

properties and uses (external)
Anti-inflammatory; anti-rheumatic; astringent; anti-oedema; helps to nourish, tone and protect vein walls.
Ideal for rheumatism, varicose veins, haemorrhoids, chilblains, bruises, and swollen ankles.

dosage
1:5 tincture of 35% alcohol.
1-2% used in creams / lotions and gels and up to 5% can be used in ointments / balms.

contraindications
Although this medicine can be taken internally it is best to seek the advice of a qualified medical herbalist before doing so as it could be toxic in the wrong dose / indication.
Note: do not apply to broken skin.

Commiphora molmol: myrrh

family
Burseraceae.

part used
Resin ('tears').

constituents
Gums and acidic polysaccharides; resin; and volatile oils (e.g. heerabolene, eugenol, and various furanosesquiterpenes).

properties and uses (external)

Antiseptic and bacteriostatic (especially against staphylococcus aureus and other gram-positive bacteria); anti-inflammatory; astringent; slightly analgesic; and vulnerary.

It's astringent and antiseptic properties make it useful to help clean, sterilise, and heal cuts, and in the treatment of acne and boils (combines well with *Echinacea purpurea / angustifolia* tincture for this) as well as mild inflammatory skin problems.

dosage

1:5 tincture of 90-96% alcohol.

1% used in creams/lotions and gels and up to 5% can be used in ointments / balms.

contraindications

Owing to the high alcohol content of this tincture, care should be taken when applying directly to the skin as it can have quite a drying action (only apply a few drops to small cuts). Otherwise it is best incorporated into other preparations for application (especially ointments / balms).

Echinacea purpurea / angustifolia: purple coneflower

family

Asteraceae (formerly *Compositae*).

part used

Whole of the plant in the case of *Echinacea purpurea*, and just the root / rhizome in the case of *Echinacea angustifolia*.

constituents

Alkamides (notably isobutylamides with olefinic and acetylenic bonds); caffeic acid esters (e.g. echinacoside and cynarin); various polysaccharides; volatile oil (e.g. humulene); echinolone; flavonoids; and betaine.

properties and uses (external)

Antimicrobial / antiseptic; immuno-stimulant; anti-inflammatory; vulnerary; and anti-allergenic.

Used in the treatment of boils; acne; abscesses; certain types of eczema; and herpes zoster (shingles).

dosage

1:5 tincture of 35% alcohol.

1-2% used in creams / lotions and gels and up to 5% can be used in ointments / balms.

contraindications

There are no known contraindications for this TR when used externally.

Glycerrhiza glabra: liquorice

family
Leguminosae.

part used
Root.

constituents
Triterpene saponins (up to 6% glycyrrhizin); flavonoids (e.g. isoflavones, liquiritin, isoliquiritin, and formononetin); polysccharides; sterols; coumarins; and asparagin.

properties and uses (external)
Anti-inflammatory; emollient; and antipruritic.
Especially useful for itchy / sore skin disorders such as eczema, psoriasis, and urticaria.

dosage
1:5 tincture of 35% alcohol.
1-2% used in creams / lotions and gels and up to 5% can be used in ointments / balms.
Note: can also be found as a fluid extract (i.e. 1:1 tincture), in which case the amount used can be lowered due to higher potency. This preparation may also give a certain amount of discolouration to the skin when applied owing to its dark colour and thick consistency. This is not permanent and usually fades within a few days.

contraindications
There are no known contraindications for this TR when used externally and at the recommended amounts above.

Stellaria media: chickweed

family
Caryophyllaceae.

part used
Aerial parts.

constituents
Triterpenoid saponins; coumarins; flavonoids; carboxylic acids; and ascorbic acid (vitamin C).

properties and uses (external)
Emollient; vulnerary; antipruritic; anti-rheumatic; and also has slight refrigerant (cooling) properties like *Mentha x piperita* (peppermint).

This has been used for hundreds of years as a remedy for chronic skin conditions, namely boils; painful eruptions; varicose veins; ulcers; eczema; urticaria; abcesses etc, as well as muscular rheumatism and inflamed (esp. gouty) joints.

dosage
1:5 tincture of 35% alcohol.
1-3% used in creams / lotions and gels and up to 6% can be used in ointments / balms.

contraindications
There are no known contraindications for this TR when used externally.

appendix 5: the chemistry of emulsification

fats and fatty acids

Officially classified as 'lipids', fats (including oils) are mostly made up of 'fatty acids' which are composed of two parts: a fatty (non-polar, water-hating) 'tail', made up of a carbon chain and ending in a methyl group, and an acid (polar, water-loving) 'head', also known as a carboxyl group (see fig 73: basic structure of a fatty acid, below).

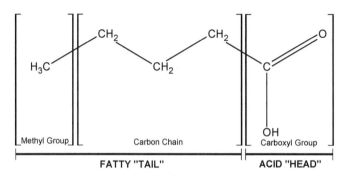

fig 73: basic structure of a fatty acid

Fatty acids are: the major building blocks of fat in organisms; an important source of energy for the body (either used directly or converted into sugars first); needed for the maintenance of healthy cells; used to make hormones and neurotransmitters essential for our general well-being.

Fatty acids can vary in length depending on how long their fatty tail is (generally the longer this is, the higher the material's boiling point will be, and so the more likely the material will be a solid at room temperature).

Fatty acids can be either saturated or unsaturated. If a fatty acid is saturated it means that there are no double bonds present in the fatty tail (see fig 74: basic structure of saturated fatty acids). As a result, these types of fatty acids bunch closer together, have higher boiling points, and are harder to break down in the body (these are mostly found in animal-sourced fats).

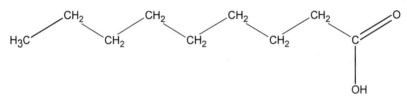

fig 74: basic structure of saturated fatty acids

If a fatty acid is unsaturated it means that that there are one or more double bonds present in the fatty tail. If there is just one double bond present, the fatty acid is known as mono-unsaturated and if more than one double bond is present it is known as polyunsaturated (see figs 75 and 76: basic structure of mono-unsaturated, and polyunsaturated fatty acids). As a result, these types of fatty acids have 'kinks' in their tails which inhibit them from bunching tightly together, allowing them to be more fluid, and easier for the body to break down.

They also contain fewer hydrogen atoms, making them lighter compared to their respective saturated cousins, resulting in lower boiling points and a greater likelihood that they will be liquids at room temperature (these types of fatty acids are mostly found from oils sourced from plant material).

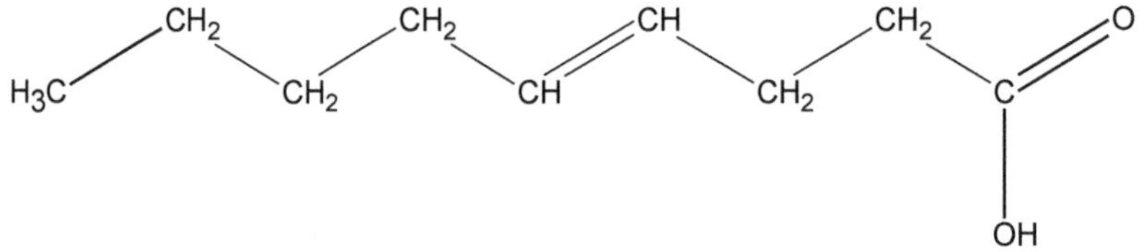

fig 75: basic structure of a mono-unsaturated fatty acid

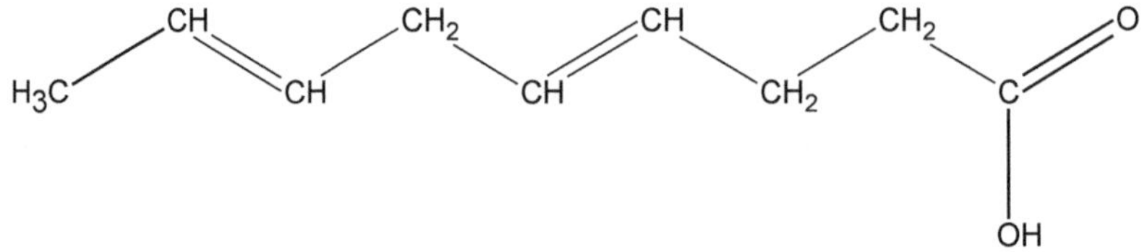

fig 76: basic structure of a polyunsaturated fatty acid

Fats and oils are made up of combinations of these fatty acids, which in turn give the substances their respective properties (including their feel, look, smell, taste, and therapeutic application).

water

This is the most abundant and widely-used solvent on earth. It is made up of three atoms - two oxygen and one hydrogen - and is given the formula H2O (see fig 77: the structural formula of water, below).

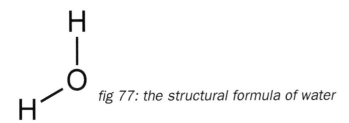

fig 77: the structural formula of water

oil and water don't mix

Everyone knows that oil and water don't mix. But why is this?

When oil is added to water, its molecules immediately separate themselves from the water molecules (usually floating on top of the water). Even if the mixture is vigorously shaken to mix the oil with the water, the two phases will mix momentarily, but when left to stand, the two will always separate out into their own phases. The long fatty tails of the fatty acids present in the oil are strongly hydrophobic (water-hating), and so in their attempt to get away from the water, they orientate themselves in an upright position so that only their acid heads (being weakly negatively charged and slightly hydrophilic, or water-loving) are in contact with the water - thereby effectively producing a fatty barrier on top of the water phase (see fig 78: water and oil don't mix!).

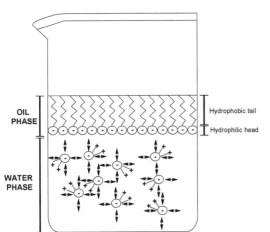

fig 78: oil and water don't mix!

A water molecule is polar and an oil molecule is non-polar. Polarity is too complicated a concept to explain adequately in this book, but here's a brief description. A polar molecule has its electrical charges arranged somewhat asymmetrically around the molecule, so that in the case of water, the hydrogen electrons are much more attracted to the oxygen nucleus than to the hydrogen nuclei, increasing the negative charge on the oxygen side of the molecule. Oils have non-polar molecules, and their electrical charges are much more evenly distributed. (see figs 79 and 80).

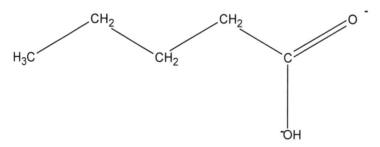

fig 79: electrical charges on fatty acids

Substances are only miscible with other substances that have similar characteristics (like dissolves like). Water is miscible with other polar compounds but not with non-polar compounds, and oil is miscible with other non-polar compounds but not with polar compounds.

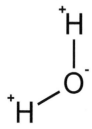

fig 80: electrical charges on water

Secondly, water molecules form extremely strong bonds with one another (called hydrogen bonds). This can be demonstrated by carefully filling a glass with water to the brim, but without spilling any. If you look closely you may be able to see that the water is bulging over the top of the glass due to its high level of surface tension. This surface tension is because of strong hydrogen bonds. These hydrogen bonds are due to the extraordinarily strong attraction that the hydrogen in water molecules have for the oxygen of other water molecules around them (see fig 78: water and oil don't mix!).

In a similar way, oil molecules also have a very strong affinity with each other. This is not due to hydrogen bonds, but to 'London forces' or 'dispersion forces', which cause the large oil molecules to bunch together (usually floating on top of the water layer).

So the water molecules are unable to hydrogen-bond with the oil molecules, and the oil molecules' attraction (via dispersion forces) to the water molecules is too weak compared to their attraction to other oil molecules. The result is a separation of the two substances.

Finally, as mentioned in the first point above, like dissolves like. The size and composition of oil molecules are very unlike the size and composition of water molecules. Whether the oil is from a vegetable, animal, mineral or synthetic source, oil molecules have little in common with water molecules and are therefore immiscible with them.

emulsifiers

Creams and lotions are mixes of water and oil - so how can we achieve this in light of the processes explained above?

As already mentioned, when the oil and water mix is shaken vigorously, the two phases will mix momentarily. This is called 'unstable emulsification', because after a while the two phases will separate out again.

In order to make these two phases mix together permanently (i.e. form a 'stable emulsification'), we need to apply other compounds to the mix which will act as mediators between the two phases.

Chemicals that do this are known as emulsifiers (also sometimes called stabilisers), and are used in many food products and cosmetics produced today.

Like fatty acids, an emulsifier consists of two parts: one hydrophilic (also known as lipophobic), and the other hydrophobic (also known as lipophilic). However, because they tend to be larger molecules, they are able to effectively act as bridges between the oil and the water.

There are two main types of emulsifiers: oil in water, and water in oil, and they are usually used in combination to balance each other's actions and produce a more stable product (see fig 81: action of emulsifiers - oil in water and water in oil).

Examples of common emulsifiers used in foods and cosmetics are: glyceryl stearate (a water in oil emulsifier), sodium stearoyl lactylate (an oil in water emulsifier) and cetearyl glucoside (which has combined water in oil / oil in water properties).

For emulsification to take place it is imperative that the right amount and type of emulsifier is used in the right phase and that they are mixed in thoroughly. It is also important that the right phase is added at the right temperature (i.e. for oil in water emulsification, the oil phase is added to the water phase, not the other way around, and this is done when both phases have reached a temperature of approximately 75-80°C).

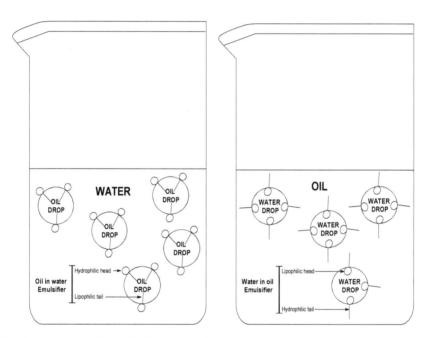

fig 81: action of emulsifiers - oil in water and water in oil

Other factors can affect the way the emulsifiers work - for example the pH of the product, or the force and duration of mixing will affect the final consistency of your end product.

appendix 6: recommended raw materials

aloe vera gel

content
Aloe barbadensis

origin
Plant source, usually refined if commercially obtained.

use
Soothing, moisturising, vulnerary.

application
Add 10-25% (reduce water phase accordingly) to lighten the consistency of heavy creams / lotions.

comments
May need to add extra moisturisers to counteract the drying effect this material can display.

cautions and contraindications
There are no known cautions or contraindications.

carbamide

content
Urea

origin
Animal source or synthetically produced.

use
Moisturiser

application
Use 3-5% as a moisturiser and antiseptic agent in foot / hand creams (can use up to 10% in foot creams to help soften the skin.

comments
Dissolves easily in water and alcohol.

cautions and contraindications
This may cause irritation to the skin with some people if used in high doses. Do not apply to broken / damaged skin.

carotene

content
ß carotene (provitamin A)

origin
Plant source (carrot, *Daucus carrota*)

use
1. antioxidant 2. colourant

application
Add 2 drops during the fat phase to avoid undesirable discolouration of product due to addition of infused waters etc.

comments
It is recommended that Vitamin E oil (tocopherol) is used in combination with this material.

cautions and contraindications
Generally thought of as being non-toxic / non-irritant. However, some people have been known to develop skin sensitivity to this material.

cellulose gum

content
Cellulose - E460-466 (hydroxyl cellulose)

origin
Plant source

use
1. thickener 2. stabiliser 3. mild moisturiser

application
Add to heated water (50°C) and whisk to produce a gel. Can also add to creams to help thicken them (use in place of xanthan gum and add to water stage before heating).

comments
Much more stable and better tolerated than other thickening agents. Also, produces a translucent rather than an opaque gel.

cautions and contraindications
There are no known cautions or contraindications.

cetearyl glucoside (also known as vegetal)

content
Cetearyl glucoside

origin
Semi–synthetically produced from plant sources (corn syrup / coconut fat)

use
Emulsifier (combined oil in water and water in oil).

application
Used as an emulsifier on its own (i.e. no need for any other emulsifiers) and without the need for thickening agents (e.g. cetyl alcohol) to make 'lighter' creams / lotions. Add 5% to the fat phase as you would normally do with glyceryl stearate.

comments
A modern emulsifier made according to ecological principles, produced without the use of chemicals or solvents. It is important that the mix is whisked gently so as not to get any air into it or it may become too runny. The product will not work if the ingredients added at the 35-40ºC stage are greater than 10%.

cautions and contraindications
At this time there are no known cautions or contraindications.

cetyl alcohol

content
Cetyl alcohol; cetanol

origin
Used to be obtained from sperm whale, but now semi-synthesised from a plant source.

use
1. thickener 2. emulsion stabiliser 3. emollient

application
Work out the % of glyceryl stearate used in your mix. Then take this figure and reduce it by 0.5-1% to get the amount of cetyl alcohol you should add (during the fat phase).

comments
It is not soluble in water, but is easily dissolved in alcohol, ether or fatty oils.
cautions and contraindications
May cause contact dermatitis in some individuals; otherwise considered low in toxicity, and safe to use.

citric acid

content
Citric acid (E330)

origin
Plant source, or synthetically produced.

use
1. acidity regulator 2. mild anti-microbial

application
Add a small amount after all other ingredients have been added (monitor pH with litmus paper) to gels, creams / lotions to ensure products maintain the natural pH of the skin, to act as a mild preservative, and to activate some preservatives (e.g. euxyl K700).

comments
Mixes easily with water and alcohol. Make sure you don't add too much as this can make the pH too low, and result in the separation of your product.

cautions and contraindications
As long as it is not used in too high a concentration, it is regarded as safe to use. As it is an acid, contact with eyes and mucus membranes to be avoided.

euxyl K700 (aka preservative K)

content
1. benzyl alcohol 2. phenoxyethanol 3. potassium sorbate

origin
Synthetically produced

use
Preservative

application
Add 0.5-1.5% to gels, creams / lotions when the product reaches 25°C. The pH is then tested and adjusted with the use of an acid until it is below pH 5.5 for this material to be effective.

comments
The product should keep for 1.5-3 years (depending on what other ingredients are present). The use of this material has recently been approved by the Soil Association for use in organic skin-care products.
cautions and contraindications
This material contains potassium sorbate and so may not be tolerated by some people (see appendix 7: raw materials to avoid).

glycerine

content
Glycerol (E422)

origin
Animal / plant sources or synthetically produced.

use
1. moisturiser 2. stabiliser

application
Add 3-5% to water phase of gels, creams / lotions to enhance moisturising effect on the skin's surface.

comments
Dissolves easily in water.

cautions and contraindications
Generally thought of as being non-toxic / non-irritant. However, it is advised that you should not use more than 15% in your base product as this can cause skin sensitivity in some individuals, and produce a drying effect on the skin (esp. if the surrounding atmosphere has low humidity). Manufacturers are not obliged to disclose the source of this material unless specifically requested – useful to know for vegans etc

glyceryl stearate

content
Glyceryl monostearate (E471)

origin
Animal source, or semi-synthesised from a plant source.

use
Emulsifier (water in oil)

application
Add 2-3% to fat phase (exact amount used depends on how big the fat phase is, e.g. must be 50-60% oil or more).

comments
Must be used in combination with sodium stearoyl lactylate and cetyl alcohol. Must be added to the fat phase not the water phase.

cautions and contraindications
Produces little or no toxicity; however, may cause skin allergies in some individuals.

honey moisturiser

content
Hydroxy-propyltrimonium honey

origin
Semi-synthesised from honey

use
Moisturiser

application
Add 2-5% to water phase of gels, creams / lotions to enhance moisturising effect on skin's surface.

comments
Twice as effective as glycerine as a surface moisturiser.

cautions and contraindications
Generally considered non-toxic / non-irritant. However, this is still a relatively new material, and so more research is needed.

lactic acid

content
Lactic acid (E270)

origin
Synthetically or semi-synthetically produced.

use
1. acidity regulator 2. mild anti-microbial

application
Same as for citric acid.

comments
Water and alcohol soluble

cautions and contraindications
Same as for citric acid.

natural moisturising factors (NMF)

content
1. sodium lactate
2. sodium pyrrolidone carboxylic acid (sodium PCA)
3. sodium phosphate

4. carbamide

5. hydrolysed vegetable protein

origin

Both synthetic and plant sources (e.g. soy bean and asparagus extracts).

use

Moisturiser

application

Add 2-5% to creams / lotions when mix has cooled below 40°C. Commonly combined with glycerine (i.e. add 2% each of natural moisturising factors and glycerine to your mix).

comments

Moisturises at a deeper level in the skin (differs from glycerine and sorbitol).

cautions and contraindications

No known cautions or contraindications.

propylene glycol

content

Propylene glycol

origin

Synthetically produced.

use

1. mild moisturiser

2. mild emulsifier

3. mild anti-microbial

application

Add to creams and lotions in small amounts in combination with other moisturisers and emulsifiers. Use instead of vegetable oils or water to make macerated / infused extracts.

comments

It mixes easily with both water and alcohol. More commonly used in the extraction of EOs and herbal medicines where water, alcohol, and oil are not suitable solvents.

cautions and contraidications

Generally thought of as being non-toxic / non-irritant, although the accepted daily intake (ADI) is 25mg per Kg of body weight. It has been used in anti-freeze, but these preparations are now more likely to contain ethylene glycol rather than propylene glycol.

sodium benzoate

content
Sodium benzoate (E211)

origin
Synthetically produced, or from a plant source.

use
Preservative

application
Add 0.2-0.5% (amount used is increased in proportion to the water content of the product, and also if any infused / macerated waters have been employed).

comments
It works best in acidic conditions (i.e. less than pH 6), so you may need to add some citric / lactic acid to the mix.

cautions and contraindications
This preservative has been well researched over many years. However, as with paraben, some people may have an allergic reaction to this product. Not as effective as paraben and so shelf life may be shorter. Not recommended for use in products containing water extracts (e.g. decocted / infused / macerated / percolated waters).

sodium stearoyl lactylate

content
Sodium stearoyl-2-lactylate (E481)

origin
Animal source, or semi-synthesised from a plant source

use
Emulsifier (oil in water)

application
Work out percentages of glyceryl stearate and cetyl alcohol in your mix. Add these two percentages together to get the percentage of sodium stearoyl lactylate that needs to be added.

comments
Must be used in combination with cetyl alcohol and glyceryl stearate to produce a stable emulsion. Must be added to water phase not fat phase.

cautions and contraindications
No known cautions or contraindications.

vitamin A

content
Retinol palmitate

origin
Synthetically produced

use
1. moisturiser
2. skin tonic

application
Add 0.1-1% to finished product in cases of ageing, dry, lifeless skin or in the treatment of acne and psoriasis.

comments
Some research suggests that the presence of this vitamin in products may help to reduce wrinkles.

cautions and contraindications
Generally thought of as being non-toxic / non-irritant when used externally.

vitamin B5

content
D-panthenol; pantothenic acid

origin
Synthetically produced

use
1. moisturiser
2. skin-tonic
3. anti-inflammatory
4. anti-microbial

application
Add 3-6% to the mix when it has cooled below 40°C for its moisturising and soothing properties.

comments
It is water soluble and very stable when exposed to light and air, but becomes unstable when heated over 50°C.

cautions and contraindications
Generally thought of as being non-toxic / non-irritant when used externally.

vitamin E

content
(and ß) tocopherol

origin
Synthetically produced or from a plant source

use
Antioxidant

application
Add 0.5-1% to fat phase when it has cooled below 40°C to help increase the shelf life of the product (i.e. stop it from going rancid). You can also add to MOs / IOs to increase their shelf life. Add 1-3% to products to give antioxidant properties to skin too.

comments
It is insoluble in water but soluble in fats / other oils. It is sensitive to oxidation and UV light, but can tolerate high temperatures (200°C). Synthetically-produced type is better suited for use as an anitoxidant, to reduce rancidity of oils in product.

cautions and contraindications
No known cautions or contraindications.

xanthan gum

content
Xanthan gum (E415)

origin
Semi-synthetically produced. Produced by the bacteria Xanthomonas compestris through the fermentation of glucose.

use
1. thickener
2. mild moisturiser

application
Add 1-2% to heated water to produce a gel. Can also add to creams to help thicken them up if too runny.

comments
Usually gives an opaque gel. If you blend too vigorously or allow air bubbles into the mix then you may end up with a product that has the consistency of chewing gum.

cautions and contraindications

No known cautions or contraindications.

Note: vegetable oils / waxes and herbal / aromatic extracts (including EOs) have been omitted from this appendix. This is because there are already many good books available on these materials, and their inclusion here would simply create an appendix too large to be practically usable.

Note: for more information on DWs see appendix 2: distillate water monographs, and for more information on TRs see appendix 5: herbal tincture monographs.

appendix 7: raw materials to avoid

2-bromo-2 nitropropane-1,3-diol

content
2-bromo-2 nitropropane-1,3-diol

origin
Synthetically produced

use
1. preservative
2. solvent

application
Used in cosmetics as a preservative. Also used as a solvent for nail varnish.

comments
Should not be used in concentrations above 0.1%.

cautions and contraindications
Found to cause allergies and contact dermatitis in some individuals. Can form nitrosamines if combined with triethanolamines (see below). Inhalation of this substance can lead to loss of appetite, diarrhoea, and headaches.

borax

content
Sodium borate

origin
Mineral source

use
1. emulsifier
2. thickener

application
Used extensively to make creams and lotions

comments
This product is easy to use as it miraculously helps blend fats with water.
cautions and contraindications

Extended use of products containing borax may result in dry, brittle skin. Some studies suggest its link to a type of poisoning that can lead to anaemia in small children. Research suggests that borax can penetrate the skin, is a powerful irritant and may even cause cancer. Use of borates has been linked to the possible development of foetal malformations.

butylated hydroxytoluene (also known as BHT)

content
Butylated hydroxytoluene

origin
Synthetically produced

use
Antioxidant

application
Added to fat-containing products to prevent them from going rancid. Also may be added to EOs to prevent them from oxidising.

comments
The research done so far seems to provide conflicting views about the safety of this substance.

cautions and contraindications
There is some evidence that suggests that this substance may be carcinogenic. It is also possible that it is easily absorbed by the skin.

collagen

content
Collagen

origin
Animal source (usually from young animal carcasses – e.g. piglets or calves). In some cases, from aborted human foetuses, or even exectued prisoners, in some countries!

use
1. skin tonic 2. mild moisturiser

application
Used extensively in 'anti-wrinkle' creams, some shampoos, and 'pout' injections.

comments
It is thought that it helps tone the skin by replacing lost natural collagen. However, it is unlikely that the collagen is even absorbed into the skin to have any effect.

cautions and contraindications
Generally thought of as being non-toxic / non-irritant when used externally.

euxyl K100 (also known as Actizide AC)

content
1. benzyl alcohol
2. methylchloroisothiazolinone
3. methylisothiazolinone

origin
Synthetically produced

use
Preservative

application
Used as a preservative against bacteria and fungi in cosmetics. Also used in anti-freeze for cars and as a wood preservative.

comments
This product (and kathon CG) is used in around 10-20% of cosmetics on the market today.

cautions and contraindications
This product is suspected to be one of the most common causes of allergic reactions to cosmetics. Symptoms include: redness, blisters, itchiness, boils, rashes, peeling, and swelling of the skin!

gelatine

content
Gelatine

origin
Animal source. Mostly from boiled bones from the slaughterhouse.

use
1. thickener
2. mild moisturiser

application

Used to make gels and thicken creams. Also used to make lozenges and capsules for medicines.

comments

This substance is still used as a thickener in a lot of food preparations today.

cautions and contraindications

Since the occurrence of CJD in the 90s, it is advised that this material not be used (either externally or internally), as the body parts used to extract this material have been shown to hold the greatest concentration of prions.

isopropyl myristate

content

Isopropyl myristate

origin

Semi-synthetically produced

use

Oil substitute

application

Commonly used in the cosmetics industry as a substitute for vegetable oils.

comments

It helps to make the skin feel soft without being oily, and does not go rancid.

cautions and contraindications

It has been shown to react with triethanolamines to create compounds that can be absorbed by the skin and possibly cause health problems.

kathon CG

content

1. methylchloroisothiazolinone
2. methylisothiazolinone

origin

Synthetically produced

use

Preservative

application

Used as a preservative against bacteria and fungi in cosmetics. Also used in anti-freeze for cars and as a wood preservative.

comments

This product (and euxyl K100) is used in around 10-20% of cosmetics on the market today.

cautions and contraindications

This product is suspected to be one of the most common causes of allergic reactions to cosmetics. Symptoms include: redness, blisters, itchiness, boils, rashes, peeling, and swelling of the skin.

paraben

content

1. phenoxyethanol
2. methylparaben
3. ethylparaben
4. propylparaben (E214-218); may also have butylparaben

origin

Synthetically produced

use

Preservative

application

Broad spectrum preservative that has been commonly used in food and cosmetic products.

comments

Has been used for the past 60 years and is generally regarded by commercial manufacturers as being well tolerated when used at recommended levels of concentration (i.e. 0.4-0.8%).

cautions and contraindications

Some individuals may have allergic reactions specific to this material.

May be weakly oestrogenic (esp. propyl and butylparaben) and should be avoided by pregnant women as a safety precaution. Recently some researchers have also detected significant levels in samples of breast tumour tissue, although further research is needed to give a clearer picture of any direct link between parabens and breast cancer.

Research has indicated this material's possible effect on the male reproductive system (in particular a decrease in both serum testosterone concentration and daily sperm production).

petroleum products

content
Mineral oils / waxes

origin
Semi-synthetically produced
use
1. emulsifier
2. solvent

application
Used in gels, ointments / balms, creams / lotions and in medicines.

comments
Commonly used as paraffin wax / oil or petroleum jelly 'ointment'.

cautions and contraindications
May cause discoloration of the skin. These products have been known to dry out the skin dramatically, may prevent absorption of vitamins, can clog pores, and may also be carcinogenic (due to the presence of impurities, e.g. polycyclic aromatic hydrocarbons).

potassium sorbate

content
Potassium sorbate (E202)

origin
Synthetically produced or from a plant source

use
Preservative

application
Used in skin care products as a preservative against yeasts, mould and certain bacteria.

comments
For this material to work it is best to increase the acidity of your product to around pH 5 or less.

cautions and contraindications
Quite a lot of people are intolerant of this material, resulting in a redness and irritation of the skin.

triethanolamine (also known as TEA)

content
Triethanolamine

origin
Synthetically produced

use
Emulsifier

application
Used in cosmetic creams and lotions

comments
It has been used for several decades and is still in use today. Why is this chemical used in cosmetics?

cautions and contraindications
This chemical has been found to severely irritate eyes and skin (esp. mucus membranes). It has also been found to be able to penetrate the skin and cause liver damage. When combined with nitrate ions (commonly found in drinking water), a chemical reaction occurs producing nitrosamine (a carcinogenic substance).

appendix 8: tincture ratio chart for 96% ethanol

water volume required (ml)	alcohol volume required required (ml)	final strength of TR required (%)
739.6	260.4	25
687.5	312.5	30
635.4	364.6	35
583.3	416.7	40
531.2	468.8	45
479.2	520.8	50
427.1	572.9	55
375.0	625.0	60
322.9	677.1	65
270.8	729.2	70
218.7	781.3	75
166.7	833.3	80
114.6	885.4	85
62.5	937.5	90

appendix 9: cream recipe variation table

raw materials	rich cream (dry skin)	regular cream (normal skin)	light cream (oily skin)	moisturiser	lotion
vegetable oil	22%	14%	11%	9%	8%
vegetable butter	3%	1%	-	-	-
cetyl alcohol	2%	2%	1.5%	2%	1%
gyceryl stearate	3%	3%	2%	2.5%	1%
sodium stearoyl lactylate	5%	4.5%	3.5%	4%	2.5%
distilled water	65%	75.5%	82%	82.5%	87.5%
total	100%	100%	100%	100%	100%

Remember that when other ingredients (either fat-soluble or water-soluble) are used, these percentages should be adjusted so that the recipes still add up to 100% (e.g if 3% DW is used in the regular cream recipe then only 72.5% distilled water is used).

Note: the above does not apply to glycerine, CO_2 extracts, florasols, aromatic / herbal extracts, resins / gums, expressed / percolated oils, absolutes, EOs, preservatives, pH stabilisers (acids), vitamins / minerals, or xanthan gum; so these are added without adjustment to the above percentages.

glossary

antipruritic a medication that reduces pruritis, or itching.

antipyretic a medication that reduces fever.

astringe draw or bind together.

bacteriostatic restricting the growth of bacteria.

cicatrisant medication that promotes the healing of a wound.

cohobation (see page 78) repeat distillation.

contraindication anything that makes the use of a medication or a medical procedure inadvisable.

emollient softening and soothing.

gram-positive bacteria bacteria that stain dark blue / violet when Gram's stain is applied to them (as apposed to a red/pink stain as displayed by gram-negative bacteria); they employ specific disease-causing exotoxins which may be part of a response to environmental stress, and are generally more dangerous than their gram-negative counterparts. Examples are: Clostridium tetani (can cause tetanus); Streptococcus pneumoniae (can cause meningitis); Staphylococcus aureus (can cause abscess formation and surgical wound infection); and Streptococcus pyogenes (can cause scarlet fever and necrotising faciitis).

haemostatic arresting blood flow or haemorrhage.

monoterpenes belong to a family of chemicals called the terpenes and are commonly found in a variety of essential oils (such as pine, cypress, and most citrus oils). They contribute little to the aroma of essential oil; nevertheless, their structure allows them to readily penetrate the skin's surface, and their presence enables other chemicals in the essential oil to do likewise (even if these chemicals are not able to do so on their own). They also display a decongestant effect by helping to break up and loosen clogged catarrh and are highly effective as air purifiers / deodorisers.

marc old / spent herb material that has been used in a distillation.

monograph condensed piece of information about a particular plant.

non-polar generally, insoluble in water.

phototoxic causing the skin to become susceptible to damage by light.

postpartum after giving birth.

reflux (see page 79) a process of evaporation using a condenser so that any vapours given off condense back to a liquid for further evaporation.

safflower oil thistle oil.

shea butter a whitish / yellowish fat from the seeds of the shea tree.

silicone grease different from silicon - silicon is the hard element; silicone is made from silicon plus other elements. It is soft, and is used as a lubricant.

Sitz bath a small bath that covers the hips and buttocks; can be warm or cold, and contain medication.

spritzer in the context of this book - a spray.

styptic a substance that stops bleeding by constricting tissues or blood vessels.

supercritical as in 'supercritical carbon dioxide' is when the gas is subjected to low temperature and / or high pressure so that its properties resemble those of a liquid; useful in extracting plant essences because it is non-toxic, and the relatively low temperatures don't denature the plant material.

tocopherol vitamin E.

venous decongestant medication that clears blood vessels.

vulnerary a remedy used in the treating or healing of wounds.

resources

For up-to-date information, courses, books, products & services, see www.lowimpact.org/essential_oils

raw materials & other supplies

- Abbey Essentials, 69 Bethel Street, Norwich, Norfolk, NR2 1NR
  01603 610883 www.abbeyessentials.co.uk
  Essential oils, information and blending equipment
- Avicenna, Bidarren, Cilcennin, Lampeter, Credigion, Wales, SA48 8RL
  +44 (0)1570 471000 avicenna@clara.co.uk
  Organic herbs, tinctures, creams, essential oils etc.
- Chesham Speciality Ingredients Ltd, Cunningham House, Westfield Lane, Harrow, HA3 9ED, UK
  +44 (0) 208907 7779 http://www.chesham-ingredients.com/
  Specialist suppliers to the UK cosmetics and toiletries industry
- Dormex Containers Ltd, Dormex Works, Chester Road, Sutton Weaver, Cheshire, WA7 3EG, UK
  +44 (0)8706 099399 www.dormex.co.uk
  Suppliers of plastic bottles and goods.
- Elixir Herbal - Botanical Pharmacy, 49 Redcliffe Walk, Aylesbury, Buckinghamshire, HP19 9LB, UK
  +44 (0)1296 481306 www.elixirherbal.com
  Herbal products, herbal / aromatic pharmacy and laboratory supplies, distillation kits (incl. extra equipment and spare parts)
- Essentially Oils Ltd, 8-10 Mount Farm, Junction Road, Churchill, Chipping Norton, Oxfordshire, OX7 6NP, UK
  +44 (0)1608 659544 www.essentiallyoils.com
  Essential oils, base oils and other aromatherapy sundries. Also provides GCMS analysis service for essential oils and herbal extracts
- Hayman Ltd, 70 Eastways Industrial Park, Witham, Essex, CM8 3YE, UK
  +44 (0)1376 517517   www.hayman.co.uk
  96% Ethanol BP by fermentation in packs of 4 x 2.5-litre bottles or 25-litre drums (also certified organic alcohol in 25-litre drums)
- Just Aromatherapy, 45 Thorpe Road, Melton Mowbray, Leicestershire, LE13 1SE   07905 565 886
  www.justaromatherapy.co.uk
  Family company supplying essential oils, plus lots of info

- Materia Aromatica, 7 Penrhyn cresent, London, SW14 7PF
  +44 (0)2083 929868   www.materia-aromatica.com
  Organic essential oils and other sundries
- Newport Industries Ltd, Head Office, 3rd Floor Merlin House, 20
  Belmont Terrace, Chiswick, London, W4 5UG
  +44 (0) 208 742 0333   http://www.newport-industries.com/
  Cosmetic raw materials in bulk quantities
- Organic Herb Trading Co, Milverton, Somerset, TA41 1NF, UK
  +44 (0)1823 401205   www.organicherbtrading.com
  Organic dried herbs, tinctures and other sundries
- Rutland Biodynamics Ltd, Town Park Farm, Brooke, Rutland, LE15 8DG
  +44 (0)1572 757440 www.rutlandbio.com
  Biodynamic dried herbs grown on their farm, organic tinctures,
  essential oils, macerated oils etc. They also have open days
- Summer Naturals, Moorland House, 211 Huddersfield Road,
  Stalybridge, Cheshire, SK15 3DY
  www.summernaturals.co.uk
  Essential oils and base ingredients
- Wains of Tunbridge Wells (Distributors) Ltd, Enterprise Centre,
  Chapman Way, North Farm Road, Tunbridge Wells, Kent, TN2 3EF
  +44 (0)1892 521666   www.wains.co.uk
  Glass and plastic containers, closures and dispensing systems

books

The books below are all available from LILI
- Dr Stephen Antczak & Gina Mae Antczak, *Cosmetics Unmasked*,
  2001, Thorsons Publishers. Details on the safety of ingredients
  commonly used in cosmetic products
- Thomas Bartram, *Encyclopedia of Herbal Medicine*, 1995 (1st
  edition), Grace Publishers. Now in its 2nd edition, this book is an
  A to Z of herbal medicine, and is a must in every western medical
  herbalist's practice
- Salvatore Battaglia, *The Complete Guide to Aromatherapy*, 1997,
  The Perfect (Aust) Pty Ltd. Now in its 2nd edition, this book is
  considered indispensable to both students and graduates of
  aromatherapy alike
- Suzanne Catty, *Hydrosols: the next aromatherapy*, 2001, Healing
  Arts Press. An excellent book for anyone interested in DWs,
  including both medicinal and culinary recipes

- Patricia Davis, *Aromatherapy: an A - Z*, 2000, The C.W. Daniel Company Ltd. This reference book is a must in every aromatherapy practice
- Tim Denny, *Field Distillation for Herbaceous Oils*, 2000, Denny, McKenzie Associates. This book explores the science and physics behind distillation
- Department of Trade, *Guidance on the Implementation of the Cosmetic Products (Safety) Regulations*, 2005. Up-to-date information on the legal regulations imposed on the UK commercial cosmetics industry
- Udo Erasmus, *Fats that Heal, Fats that Kill*, 1993, Alive Books. Explains what makes up the oils we consume and gives details of how some of these oils may be contributing to common health complaints.
- William Evans, *Trease and Evans Pharmacognosy* (14th edition), 1996, WB Saunders Co. Ltd. The textbook for correct identification of herbal matter for use as medicine.
- Gerald Hasenhuettl and Richard Hartel, *Food Emulsifiers and their Applications*, 1997, Chapman and Hall. Detailed information on the process of emulsification and on food grade emulsifiers
- Jan Kusmirek, *Liquid Sunshine: vegetable oils for aromatherapy*, 2002, Floramicus. A great resource for vegetable / base oils
- Donna Maria, *Making Aromatherapy Creams & Lotions: 101 natural formulas to revitalize and nourish your skin*, 2000, Storey Books. Simple recipes for home-made creams and lotions
- Gabriel Mojay, *Aromatherapy for Healing the Spirit: a guide to restoring emotional and mental balance through essential oils*, 1997, Healing Arts Press
- Mike Nixon & Mike McCaw, *The Compleat Distiller*, 2001, The Amphora Society. A must for anyone interested in more in-depth information on distillation (although it is more directed to the distillation of alcohol - which is permitted in New Zealand without a licence).
- Kurt Saxon, *Grandad's Wonderful Book of Chemistry*, 2000 (8th edition), Atlan Formularies. A fantastic collection of old formularies that provide details from setting up a small lab to courses on glassblowing
- Brian Furniss, Antony Hannaford, Peter Smith, Austin Tatchell, Arthur Vogel, *Vogel's Textbook of Practical Organic Chemistry* (5th Ed), 1989, Longman Scientific and Technical. Very detailed book on organic chemistry and correct procedures for setting up and looking after lab glassware and equipment.

- Ruth Winter, *A Consumer's Dictionary of Cosmetic Ingredients: complete information about the harmful and desirable ingredients in cosmetics and cosmeceuticals*, 1999 (5th edition), Three Rivers Press. An A – Z of commonly used cosmetic ingredients
- James Zubrick, *The Organic Chem Lab Survival Manual: a student's guide to techniques*, 2004 (6th edition), John Wiley & Sons. A brilliant book for beginners interested in applying laboratory techniques (such as distillation / herbal extraction)

websites

- www.aromacaring.co.uk/safety1.htm - Aromacaring: essential oils safety info
- http://lorien.ncl.ac.uk/ming/distil/distil0.htm - Distillation, an Introduction by M. Tham
- http://ull.chemistry.uakron.edu/chemsep/distillation - Distillation Theory and Practice
- www.levity.com/alchemy/jfren_ar.html - Alchemy by Adam McLean
- www.herbs.org - The Herb Research Foundation
- www.herbsociety.co.uk - The Herb Society
- http://world.std.com/~krahe - Herbs & Aromas: information about herbs and their essential oils, decoctions, syrups, tinctures etc.
- http://homedistiller.org - home distillation of alcohol
- http://12121.hostinguk.com/Guide%20Oils.htm - Magic Hands Guide: alphabetical guide to lots of essential and natural oils
- www.oilganic.com - essential and fragrance oil information and distillation
- www.oils4life.co.uk - Oils4Life: essential oils for sale online
- www.fromnaturewithlove.com/recipe - Recipe Database: lots of recipes for home-made cosmetics and toiletries
- www.soilassociation.org/web/sa/saweb.nsf/848d689047cb466780 56a6b00298980/01b0dc13bdaff03980256e8d00331a05!Open Document - sorry about the length of that web address, but it's the Soil Association, with a list of places to find certified organic essential oils and other health and beauty products

professional bodies

- National Institute of Medical Herbalists (NIMH): Elm House, 54 Mary Arches street, Exeter, Devon, EX4 3BA, UK
+44 (0)1392 426022   www.nimh.org.uk
Information on courses and contact details of qualified / practising medical herbalists both in the UK and abroad
- International Federation of Professional Aromatherapists (IFPA): 82 Ashby Road, Hinckley, Leicestershire, LE10 1SN, UK
+44 (0)1455 637987   www.ifparoma.org
Information on courses and contact details of qualified / practising aromatherapists both in the UK and abroad
- Aromatherapy Organisations Council: www.aocuk.net - main umbrella organisation ensuring quality in essential oil training and practice
- Aromatherapy Trade Council: www.a-t-c.org.uk - regulatory body for the UK essential oils trade

other services

- Alcohol Licence:
Obtain copies of Notice 47 and Form EX240 from your local Customs and Excise Office, or visit www.hmrc.gov.uk and enter Notice 47 and EX240 into their search facility
- Cosmetic raw material safety info:
Cosmetic Ingredient Review   www.cir-safety.org
Independent research and safety testing of commonly-used raw materials in cosmetics for safety
- Good Manufacturing Practice info:
Medical and Healthcare Products Regulatory Agency (MHRA)visit their website www.mhra.gov.uk and put good manufacturing practice into their site search facility
- Laboratory Testing of Products:
Pattinson Scientific Services Ltd, Scott House, Penn Street, Newcastle-upon-Tyne, NE4 7BG
+44 (0)1912 261300 pattinsonlab@btconnect.com

other LILI publications

Learn how to heat your space and water using a renewable, carbon-neutral resource – wood.
This book includes everything you need to know, from planning your system, choosing, sizing, installing and making a stove, chainsaw use, basic forestry, health and safety, chimneys, pellet and woodchip stoves The second edition has been expanded to reflect improvements in wood-fuelled appliances and the author's own recent experience of installing and using an automatic biomass system.

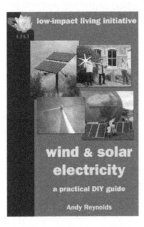

The author has been providing his own electricity from the sun and the wind for many years and in the first edition of wind and solar electricity he shared his knowledge and experience to explain how his readers could do the same.
Subsequent developments in the associated technology and UK government incentives have led him to make substantial revisions and additions to the original text, including new illustrations and photographs, for this second edition. He provides practical, hands-on advice on all aspects of setting up and keeping a home-generation system running and the text reflects his own recent experience.

The author grew up in Jamaica and was taught to make soaps by her grandmother. They grew all the plants they needed to scent and colour their soaps and even used wood ash from the stove to make caustic potash.
Her book is intended for beginners, includes both hot- and cold-process soap making, with careful step-by-step instructions, extensive bar, liquid and cream soap recipes, full details of equipment, a rebatching chapter plus information on the legislation and regulations for selling soap.
Now also available as part of an an online course at: http://lowimpact.org/online_courses_natural_soaps.html

how to spin: just about anything is a wide-ranging introduction to an ancient craft which has very contemporary applications.
It tells you all you need to know about the available tools, from hand spindles to spinning wheels, what to do to start spinning, with illustrated, step-by-step instructions, and a comprehensive guide to the many fibres you can use to make beautiful yarns.

Janet Renouf-Miller is a registered teacher with the Association of Weavers, Spinners and Dyers, and has taught at their renowned Summer School.

solar hot water: choosing, fitting and using a system provides detailed information about solar-heated water systems and is particularly applicable to domestic dwellings in the UK.
Lee Rose has 10 years of experience and involvement in every aaspect of the solar thermal industry in the UK and around the world. His book provides a comprehensive introduction to every aspect of solar hot water: including all relevant equipment, components, system design and installation and even how to build your own solar panels.

Compost toilets reduce water usage, prevent pollution and produce fertiliser from a waste product. Built properly they can be attractive, family friendly and low maintenance.
This DIY guide contains everything you need to know about building a compost toilet, plus proprietary models, decomposition, pathogens and hygiene, use and maintenance, environmental benefits, troubleshooting and further resources.

herbal remedies: how to make, use and grow them teaches you to identify, grow and harvest medicinal plants. It shows you how to make a range of simple medicines; there are sections on body systems, explaining which herbs are useful for a range of ailments, and detailed herb monographs. This second edition has been revised to take account of recent changes in UK legislation.

Sorrell Robbins is a highly-qualified, leading expert in natural health with over 15 years experience. She teaches at all levels from beginner to advanced postgraduate and is a regular contributor to many natural health publications.

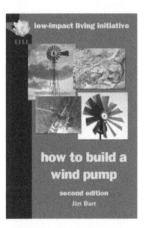

Good for developed or developing countries, the wind pump described in this book can pump rainwater, greywater, river, pond or well water for irrigation, aerate a fish pond, run a water feature or even be a bird scarer. This system does not generate electricity. The turbine is 700mm diameter and the turbine head plus rotor weighs less than 4kg. In a light-to-moderate wind it should pump about 1000 litres a day with a head of 4-5 metres. If you have good engineering skills and equipment you can fabricate nearly all of the system yourself; if you get all the parts manufactured, it's not much more complicated than DIY flatpack furniture.

This book contains 50 practical ideas for ways that you can help to stem the tide of destruction that is overtaking the ecology of our one-and-only planet. It's a random selection from the 170 topics on our website. Each chapter is a topic from one of the following five categories:

- shelter
- lifestyle
- nature
- land
- food & drink

Each topic is then divided into three sections:

- what is it?
- benefits
- what can I do?

Back in our cave-dwelling days, food smoking was used to preserve food and then our ancestors discovered just how great it makes food taste. Turan T. Turan has been a passionate smoker of food for many years, teaches courses all around UK and now crystallises his knowledge in food smoking: a practical guide. He explains the basics of cold and hot smoking; delves into the principles of combustion (he's a career fireman in another part of his life so he should know!) and explains brining and dry salt curing. He outlines how to source wood for smoking and provides plans for building a cold smoker and smoke generators. He simplifies and demystifies the process of smoking food to enable you to produce wonderful smoked food in a sustainable, eco-friendly way. Enjoy!

notes

CPSIA information can be obtained
at www.ICGtesting.com
Printed in the USA
LVHW081037120222
710988LV00008B/459

9 780956 675156